energy analysis

energy analysis

edited by
John A.G. Thomas

A collated reprint of articles from the
journals *Energy Policy* and *Food Policy*

ipc science and technology press
and
Westview Press

Energy Policy and *Food Policy* are quarterly journals published by IPC Science and Technology Press Limited.

Published 1977 in England by IPC Science and Technology Press Limited, IPC House, 32 High Street, Guildford, Surrey, England GU1 3EW

Published 1977 in the United States of America by Westview Press Inc., 1898 Flatiron Court, Boulder, Colorado 80301, USA
Frederick A. Praeger, Publisher and Editorial Director

© IPC Business Press Limited 1977. Chapter 6 is U.K. Crown copyright © 1974, and Chapter 11 is copyright © the United Kingdom Energy Authority 1975 and 1976.

All rights reserved. No part of this publication may be reproduced, stored in a retrieval system or transmitted, in any form or by any means, electronic, mechanical, photocopying, recording or otherwise, without the prior permission of the copyright holder, application for which should be addressed to either publisher.

Library of Congress Cataloging in Publication Data

Main entry under title:

Energy analysis.

1. Power resources – Addresses, essays, lectures.
2. System analysis – Addresses, essays, lectures.
I. Thomas, John Albert Gay.
TJ163.2.E456   333.7   77-4931
ISBN 0-902852-60-4 (IPC Science and Technology Press)
ISBN 0-89158-813-2 (Westview Press)

Photoset by Specialised Offset Services Limited, Liverpool
Printed in England by Kingprint Ltd, Richmond, Surrey

Contents

Introduction		vii
1	*Energy costs: a review of methods* Peter F. Chapman	1
2	*The energy cost of fuels* Peter F. Chapman, Gerald Leach and M. Slesser	14
3	*The energy costs of materials* Peter F. Chapman	27
4	*An international comparison of polymers and their alternatives* R. Stephen Berry, Thomas V. Long II and Hiro Makino	38
5	*Energy and food production* Gerald Leach	50
6	*The energy cost of goods and services: an input-output analysis for the USA, 1963* David J. Wright	62
7	*The energy cost of goods and services: an input-output analysis for the USA, 1963 and 1967* Clark W. Bullard III and Robert A. Herendeen	71
8	*The energy cost of goods and services: an input-output analysis for the Federal Republic of Germany, 1967* Richard V. Denton	82
9	*Energy analysis of nuclear power stations* Peter F. Chapman	88
10	*Nuclear power and oil imports: a look at the energy balance* J.H. Hollomon, B. Raz and R. Treitel	102
11	*Energy analysis of a power generating system* K.M. Hill, F.J. Walford and R.S. Atherton	109
12	*The economics of energy analysis* Michael Webb and David Pearce	126
13	*The economics of energy analysis reconsidered* Michael Common	140
14	*Net energy analysis – is it any use?* Gerald Leach	148
List of authors		161

Introduction

There is nothing new about analysing energy flows in a society and measuring the energy costs of goods and services. Gerald Foley in his book *The energy question*[1] quotes from H.G. Wells's *The world set free*:[2]

> Ultimately the government ... fixed a certain number of units of energy as the value of a gold sovereign ... and undertook, under various qualifications and conditions, to deliver energy upon demand as payment for every sovereign presented.

Here Wells was envisaging a society where currency was linked to physically measurable energy units, which were presumed to give an absolute scale of values to goods and services, according to the energy necessary for their manufacture and distribution. An unease that economics has not yet devised a satisfactory way of dealing with resource depletion was one of the spurs that drove many physical scientists to propose alternative ways of viewing society. The fragile supports upon which modern society rested had been recognised as long ago as 1912 by the Nobel-prize-winning physicist Frederick Soddy.[3] He stressed the central importance of energy in the society and foresaw the need to seek income sources of energy:

> Civilisation as it is at present, even on the purely physical side, is not a continuous self-supporting movement ... It becomes possible only after an age-long accumulation of energy, by the supplementing of income out of capital. Its appetite increases by what it feeds on. It reaps what it has not sown and exhausts, so far, without replenishing. Its raw material is energy and its product is knowledge. The only knowledge which will justify its existence and postpone the day of reckoning is the knowledge that will replenish rather than diminish its limited resources.

The present collection of papers arose out of a series of articles devised and designed for the journal *Energy Policy* by Peter Chapman, Gerald Leach and the editor.* The idea of the series was bred out of a small, informal meeting of British natural scientists – biologists, chemists, physicists – in London in 1973. It ran in consecutive issues of the journal from June 1974 onwards under the series title 'Energy Budgets' and generated such a large amount of interest that the December 1975 issue was wholly devoted to the subject, which had developed greatly over that time and had become more generally known as Energy Analysis. The table below, taken from Chapter 11 by Hill, Walford and Atherton, shows succinctly the various approaches to energy analysis, most of which are covered in this book.

All these papers plus two others are collected here. Chapter 5, 'Energy and food production' by Gerald Leach, is reprinted from the first issue of *Food Policy*, a sister journal to *Energy Policy*, and is itself based on a seminal work which has been published in book form.[4] The other chapter, 'The economics of energy analysis reconsidered' by Michael Common, is a lusty response by one economist to the severe criticisms of energy analysis made by two other economists, Webb and Pearce, in Chapter 12. This was published in the June 1976 issue of *Energy Policy*.

---

[1] G. Foley, *The energy question*, Penguin, UK, 1976.
[2] H.G. Wells, *The world set free*, first edition, 1914; Collins, Glasgow, UK, 1956.
[3] F. Soddy, *Matter and energy*, Williams & Norgate, London, UK, 1912: quoted in reference 1.
[4] G. Leach, *Energy and food production*, IPC Science and Technology Press, Guildford, UK, 1976.

* The inspiration and hard work of Peter Chapman and Gerald Leach is gratefully acknowledged. The enthusiasm of all the other authors must be remarked upon as well, since they all showed a high willingness to write (and rewrite!) on their various subjects.

Forms of Energy Analysis

Form of energy analysis	Insight desired	Users	Basis of conventions*
Fossil fuel accounting i.e. analysis in units of fossil fuel consumed	Resources depletion, environmental impacts	Environmentalists, resource groups etc.	Result expressed in terms of fossil fuel utilisation. All energy utilisation/conversion systems regarded as devices for burning fossil fuels. No intrinsic value attached to energy.
Fuel demand/production analysis	Utilisation patterns, changes in demand patterns, production capacity required.	Fuel and energy suppliers, planners	Economic and technical conventions
Energy systems analysis (including process analysis)	Effectiveness of utilisation of energy/free energy.	Energy technologists	Standard thermodynamic conventions
Thrift analysis	Wastage of energy	Civil servants	
Energy input/ouput analysis	Energy content of goods, use of energy in the economy	Service to other users, policy analysts	Economic conventions associated with matrix techniques
Energy flow analysis	Comparison of energy systems using different fuels	Systems analysts	Separation of all forms of energy

* All present forms of Energy Analysis ignore the energy content of labour.

This book is not meant to be a highly condensed review of energy analysis. It is a snapshot of ongoing research, a record of the growing pains of a fledgling subject. As with any new area of study, there are discourses and digressions on methodology: at times impossible claims are made for energy analysis, and at other times there is a desire to 'throw the baby out with the bathwater'.

Since this is a record of research conducted by different research groups in the UK, the USA and the Federal Republic of Germany (often completely independently), there are areas of overlap. For example, the work of Wright in the UK (Chapter 6) and Bullard and Herendeen (Chapter 7) in the USA was carried out in parallel. The latter authors, whose work became known to the editor later, have added to their own findings a valuable critique of Wright's research that exposes the sensitivity of end results in this subject to the assumptions made. Again, the three chapters on nuclear power should be read as a group since they illustrate dramatically how the end results depend heavily on the conventions adopted. To quote from Hill and co-workers in Chapter 11:

The casual observer of the protracted debate on the energy analysis of nuclear power can be forgiven for being confused. He may have heard, or read, that the payback period for nuclear power is 2-3 months from start up, or 8-24 months, or that payback occurs after 4 years, 9 years, 15 years and even never. We can add to the confusion by demonstrating that all the above answers can be derived from an agreed single set of input data merely by adopting different definitions for nuclear power (ie, by referring to a single reactor or to an extended building programme), by adopting different conventions (ie, by excluding or including transmission losses, by the use of an opportunity cost for fossil inputs or by assuming a one-to-one equivalence for thermal and electrical energy), and by adopting different rates of growth of planned output.

Very few energy analysts would call for the energy theory of value conjured up by H.G. Wells for his new society; it is neither realistic nor helpful to consider every joule of energy as being equally useful. Exposing the energy cost of beef, polypropylene, oil shale production or nuclear power growth is a valuable and in many cases an essential service, but the decisions that have to be made when all the data have been collected remain matters of social priorities and economic possibilities.

The USA now has Public Law 93-577, which requires a mandatory net energy analysis to be carried out on new energy technology developments. Gerald Leach in the final chapter casts doubt upon the usefulness of this law and of *net* energy analysis — that is, the energy cost of winning energy. He is not happy with 'the quest for long term "points of futility" and ultimate sustainable limits to human activities' that underlies many NEA and EA studies. He sees the chance of today's detailed sums being invalidated by tomorrow's technological advance as too great to make the exercise worthwhile.

Leach carefully treads a path between despair and technological euphoria and calls for a certain humility among energy analysts with the words:

The future is opaque, a dark mirror, and no less to energy analysts than to the rest of mankind. Ultimate limits can wait on more urgent and closer concerns.

JOHN A.G. THOMAS

1. Energy costs: a review of methods

Peter F. Chapman

The energy cost of a product is a deceptively simple concept: it can vary widely according to how it is calculated. This chapter explains the origin of the variations in results obtained by different workers and examines the aims of various investigations. There is no 'correct' way of apportioning energy costs or of choosing the boundary that defines the sub-system being studied, but the methods adopted must be consistent with the overall aims of the analysis. Three types of method currently in use are reviewed by the author: statistical analysis, input-output table analysis, and process analysis. Examples are given of their application to copper smelting, electricity supply, oil refining and aluminium production. A bibliography of some of the important literature is given.

Over the past two years there has been a growing realisation that the financial costs of materials and products do not provide an adequate description of the resources needed for their production. When there are no shortages of any inputs to the production system financial analysis provides a convenient decision-making framework. However, if one input does become scarce, then the implicit assumption of substitutability, inherent in financial systems analysis, leads to false conclusions. For a wide variety of reasons a number of investigators have focused their attention on the physical inputs, such as tons of steel and kWh of electricity, needed to make particular products. The forecasts of energy shortages coupled with the realisation that energy is an essential input to *all* production processes have concentrated attention on the energy inputs to, or energy cost of, various products.

At the present time there are almost as many methods of evaluating the energy cost of a product as there are workers in the field. Where, by chance, the same product has been analysed by different methods the results often vary widely. The purpose of this review of methods is to explain the origin of the variations in results so that they can be interpreted and used correctly. To accomplish this it is necessary to examine the aims of various investigations since this explains many of the assumptions made. Finally, it is necessary to show how differing assumptions and methods can account for the divergent results.

The nature of the problem
A modern industrial system, exemplified by countries such as the UK and the USA, is a complex interconnected system with many inputs and outputs. These highly developed systems are linked together, and to the so-called underdeveloped industrial systems, by flows of commodities in international trade. In some respects the total global system is a closed system. All that man's activities accomplish is a temporary change from a stock of raw material or flow of solar energy, into products such as automobiles and food which, in time, become discarded materials and dissipated energy. This view of the world – the 'spaceship Earth' concept[1] – has

---
[1] K.E. Boulding, *Economics as a Science* (McGraw Hill, 1970), chapter 2

Energy costs: a review of methods

focused attention on the depletion of non-renewable stocks, particularly fossil fuels. Many analyses of energy costs aim to evaluate the quantity of fossil fuel energy required to produce a consumer product such as an automobile or a loaf of bread.

The production of a consumer product in the UK requires inputs from all the production processes in the country and, through international trade, from all the production processes in the world. For example, a loaf of bread requires wheat which has to be milled, cooked and transported. Transport requires fuel and vehicles, for which steel, rubber, copper and energy for fabrication are necessary. Shops and bakeries need bricks, steel, cement, wood and glass; wheat production must have tractors, fertilisers, insecticides etc. It is clearly impossible to determine the proportion of all the production processes in the world needed to produce a loaf of bread, or any other single product. Any analysis must be based on a sub-system of the world, a sub-system for which all the inputs and outputs are known. The choice of sub-system is the first crucial step in evaluating an energy cost.

Three simple sub-systems of the production of a loaf of bread are shown in Figure 1. The first is confined to the bakery and the

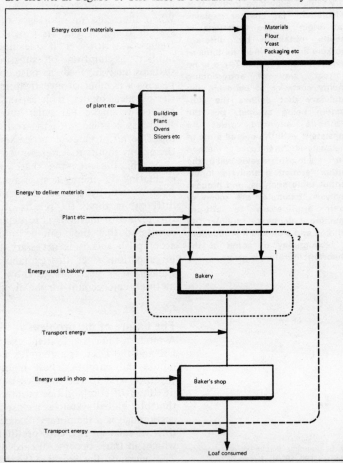

Figure 1. Possible sub-systems associated with the production of a loaf of bread. System 1 is denoted by dotted boundary, system 2 by dashed boundary, system 3 is the entire diagram.

energy cost per loaf is the energy delivered to the bakery divided by the number of loaves produced. The second sub-system includes the baker's shop. The total energy cost is:

$$\frac{\text{energy used at bakery}}{\text{loaves baked}} + \frac{\text{transport energy}}{\text{loaves delivered}} + \frac{\text{energy used by shop}}{\text{loaves sold}}$$

The third sub-system is the entire diagram and includes eight energy inputs. As the sub-system is made larger the total energy cost continues to increase. However, in a finite time it is not possible to take into account all the production processes in the world. A more feasible objective is to follow each network of inputs back from the final product until it is found that the addition of the next input makes an acceptably small difference to the total energy cost.

The choice of sub-system is one type of problem in evaluating energy costs. Another is associated with the types of energy included in the analysis and how these different energies are added together. The largest global source of energy is solar energy, yet this is usually excluded from energy costs. The production and delivery of fossil fuels involves energy consumption which may or may not be incorporated into the energy analysis. Producing secondary energy supplies, such as electricity, town gas and coke, wastes some of the energy available in primary fuels. This inefficiency may or may not be included. Most energy analyses ignore the energy input in the form of manpower or the calorific value of food. These difficulties are compounded by the various calorific values of different primary fuels and by the special role played by electricity in many industrial systems.

A third type of problem which arises is in apportioning energy costs between different products. For example many chemical processes produce two or more products in a single plant from a single set of inputs. On what basis is the energy of the plant and inputs to be divided between the products produced? On a larger scale there is the problem of apportioning the energy costs of general services, such as roads, between many users. As with all the other problems outlined in this section there is no 'correct' solution. These are not questions of fact, but of setting up the most satisfactory conventions. Analyses based on different conventions will imply different procedures for dealing with the problems.

Aims of energy studies

The fact that there is no 'correct' way of apportioning energy costs or choosing a sub-system does not mean that these are arbitrary decisions. The methods adopted within a particular analysis should be consistent with the overall aims of the analysis. Thus the first step in assessing the techniques employed in a particular study is to establish the aims of that study. Although not usually stated explicitly the aims of most studies can be inferred from published reports. There appear to be four types of aim:

1. to analyse particular processes in detail so as to deduce an energy efficiency and hence make recommendations for conserving energy;
2. to analyse the consumption of energy on a large scale either to forecast energy demand or to point to policies which could reduce future demand;
3. to analyse the energy consumption of basic technologies such as food production and mineral extraction so as to show some of the future consequences of technological trends or an energy shortage;
4. to construct energy costs and examine energy flows so as to understand the thermodynamics of an industrial system. This type of long-range aim may be coupled to projects such as 'world modelling' based on physical rather than monetary flows.

This general classification of aims is neither exclusive nor inclusive. The aims are listed hierarchically so that a study under aim (1) could actually be part of an overall project with aims (2), (3) or (4). There are, no doubt, other aims not falling readily under any of these headings.

Studies under aim (1) are often carried out by particular industries in order to make financial savings. The detailed study of particular processes requires data not normally published or generally available. An example of this type of study done outside industry is the examination of packaging.[2,3,20] By far the most popular type of study is associated with aim (2) since this corresponds most closely with 'energy policy' and the most obvious problems of the 'energy crisis'. Such studies[4,5,6] are usually based upon published national statistics. Investigations with aim (3) are often aimed at areas, such as food production[7,8] and mineral resources,[9] where conventional economics is in conflict with the predictions of 'conservationists'. In these examples an energy approach throws a new light on complex problems. The investigations can be based on published data since great accuracy is not important and the conclusions are in terms of national or global averages. Only a few studies[10,11] are associated with aim (4) on its own; however if aim (4) is the overall project aim then it has considerable influence on the assumptions made (as shown in a later section).

Methods

The implications of adopting and choosing different aims and conventions can best be illustrated by considering detailed examples. Before examining the results obtained by various authors it is necessary to outline the methods they have used. The fundamental principle of energy costing is that for a given industry or sub-system the total energy cost of all the inputs should equal the total energy cost of all the outputs. Thus if it requires 10 tons of steel (at 9940 kWh/ton steel) and 5 gallons of fuel oil (at 55 kWh/gallon) to make 10 girders, the energy cost per girder is [(10 x 9940) + (5 x 55)] ÷ 10 = 9967·5 kWh. The methods used for calculating energy costs of products differ in their sources of data

[2] I. Bousted, *Journal of the Society of Dairy Technology* Vol 27, No 3, 1974, p. 159.
[3] B.M. Hannon, *Environment* Vol 14, No 2, 1972, p. 11
[4] A.B. Makhijani, and A.J. Lichtenberg, *Environment* Vol.14, No 5, 1972, p. 10
[5] C.A. Berg, *Science* Vol 18, July 1973, p. 128
[6] E. Hirst and J.C. Moyers, *Science* Vol 179 No 4080 1973, p. 1299
[7] D. Pimental et al, *Science* Vol 182 November 1973, p. 443
[8] G. Leach and M. Slesser, 'Energy equivalents of network inputs to food producing processes' Strathclyde University, Glasgow, 1973
[9] P.F. Chapman, *Metals and Materials* Feb. 1974
[10] P.F. Chapman, 'Energy and world modelling', Seminar report, Open University 1973.
[11] D.J. Wright, 'Calculating energy requirements of commodities from the input/output table'. Paper presented at Conference, Imperial College, London, July 1973.

and techniques for deducing results. There are three types of method currently in use.

Statistical analysis
The supply of energy to various industries is available, for most industrial nations, in statistical publications such as the UK *Report on the Census of Production, 1968*. This information, coupled with data on industrial output, allows an estimate to be made of the energy cost per unit of output. For example the UK *Digest of Energy Statistics* gives the energy supplied to the iron and steel industry (1968) as 6871×10^6 therms. The output of crude steel (1968) is given as $25 \cdot 86 \times 10^6$ tons (*Iron and Steel Industry Annual Statistics*). This gives a value of $265 \cdot 7$ therms/ton steel.

This result is not a useful value for a number of reasons. The method has made no allowance for:

- the energy cost in generating the electricity and coke consumed. The 6871×10^6 therms is the energy actually delivered to the industry; it requires about 8700×10^6 therms of primary fuel consumption;
- energy sales by the iron and steel industry. Sales of gas and electricity in 1968 were 48×10^6 therms;
- energy expenditures associated with the consumption of raw materials, the depreciation of plant or the delivery of materials and products.

However all these objections can be taken into account by digging a little deeper into the published statistics. In general this method can provide an order of magnitude estimate of the energy cost of products classified by industry. It cannot take into account all the subsidiary energy costs; nor can it distinguish in detail between different products of the same industry. Since this method relies upon national statistics the sub-system assumed is the nation.

Input-output table analysis
The input-output (I/O) table of a national economy is a square matrix, A, summarising the commodities necessary to make other commodities. Thus a single entry in the table, A_{ij}, in the ith row and jth column, indicates the amount (measured in money) of commodity i required as a direct input, to produce £1.00 worth of commodity j. Thus all the inputs necessary to make £1.00 worth of commodity j are the items in the jth column of the square array.

For a given set of outputs, denoted by a vector x, the direct inputs required, denoted by a vector y, can be found by multiplying x by the matrix (the I/O table) A:

$$y = A x$$

To find the commodities, z, needed to produce the commodities y, the same procedure is adopted. Hence

$$z = A y = A (A x) = A^2 x$$

Thus all the inputs, direct and indirect, required to produce the outputs x are $A x + A^2 x + A^3 x + \ldots$ This series can be summed mathematically. The result of this analysis is thus a list of *all* the commodities required, within the nation covered by the I/O table, to produce a specified output. Clearly this method is taking a national sub-system and evaluating *all* the inputs within that system.[11-13]

There are some obvious disadvantages to this approach. Clearly the I/O table cannot be broken down into individual firms; it has to deal with industries in groups. Another disadvantage is that the method deals with transactions in financial terms, not in terms of physical quantities. This can lead to errors if commodities are liable to large price fluctuations or if some purchasers can obtain special prices for the commodity.

Process analysis
Process analysis involves three stages. The first is to identify the network of processes which contribute to a final product, as illustrated in Figure 1. Next each process within the network has to be analysed in order to identify the inputs, in the form of equipment, materials and energy. Finally an energy value has to be assigned to each input.

There are two clear problems with this method. The first is choosing an appropriate sub-system, the other is attaching energy values to particular inputs. This latter problem is crucial and can be illustrated by a simple example. The production of steel requires machines with a finite lifetime. Thus the energy cost of each machine must be averaged over all the steel processed by the machine. The machine will probably contain a great deal of steel and will have been made by other steel-containing machines. So to find the energy cost of the machine (necessary to find the energy of steel production) it is necessary to have an energy cost of steel! In practice this problem is solved by starting with an approximate energy cost of steel,[9,10] (deduced by one of the methods above) and to use this to calculate the energy of the machine and hence a better value of the energy cost of steel. In most cases this feedback interaction only contributes a small percentage to the final energy cost estimate, so provided the final result is not wildly different from the starting value it is only necessary to go round this loop once. For industries which are strongly linked, ie where a significant fraction of each one's output goes into the other, the most direct way to solve the problem is to solve the simultaneous equations involved (shown in Figure 2).

Results
The following examples of results are intended to show the care required in interpreting bald figures. In all cases the aim is not to show that one result is 'wrong' and another 'right' but simply to underline the distinctions drawn in the previous three sections.

Copper smelting
This provides a convenient example of a detailed process analysis and the variation in results due to different choice of sub-sytem. There are various kinds of smelting furnace available but recently an electric furnace has been recommended to the industry on the

[12] W.A. Reardon, 'An input/output analysis of energy use changes 1974-1975 and 1958-1963' Battelle Northwest Labs, 1971

[13] E. Hirst and R. Herendeen, 'Total energy demand for automobiles', Society of Automotive Engineers Inc., publ. 730065. 1973.

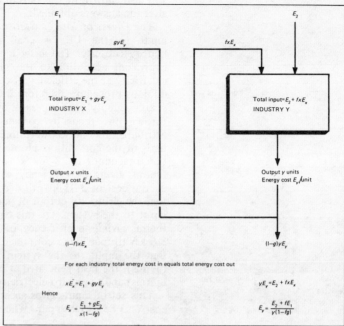

Figure 2. Two industries, X and Y, supplying each other with raw materials.

basis of better thermal efficiency. A detailed examination of the heat processes within an electric furnace[14] shows it has a thermal efficiency of 61% compared to fuel-heated furnaces with an efficiency of 27%. A comparison of the heat inputs required per ton of copper thus shows a factor of two in favour of electricity. Whether this represents a financial saving to a producer depends upon the relative prices (£ per unit of heat) of electricity and other fuels. However, as far as the industry is concerned, this is a significant energy saving.[15]

If the sub-system considered is enlarged to include the electricity supply industry and other inputs to the electric furnace the opposite conclusion results. The present efficiency of electricity generation in the UK is about 25% (see below) indicating that the supply of 1 kWh-electrical (kWhe) requires an input of 4 kWh-thermal (kWhth). Thus the 2 to 1 ratio in favour of electric furnaces becomes almost a 2 to 1 ratio against. A detailed study of both smelting systems[16] shows that for a fuel-fired system the energy cost is about 5400 kWhth/ton copper and for an electric furnace about 8000 kWhth/ton copper. It is a disturbing conclusion that in good faith an industry could improve its own thermal efficiency whilst increasing the national energy consumption.

Supply of electricity

The contradictory results obtained from two analyses of copper smelters hinged on the distinction between energy delivered (as electricity) and total energy input (to a nation). It is worth exploring this topic further if only because different authors use different conversion factors and in any case the efficiency of electricity generation is likely to change with time. This example

---

[14] O. Barth, in *Extractive metallurgy of Cu, Ni and Co* ed P. Queneau (NY, Interscience, 1960) page 251.

[15] D.G. Treilhard, 'Copper state of the Art' *Chemical Engineering*, April 1973

[16] P.F. Chapman, 'The energy cost of producing copper and aluminium from primary ore', Report ERG001, Open University 1973

Energy costs: a review of methods

will also show how the aims of a particular study can dramatically alter the answers obtained..

The *Digest of Energy Statistics* for the UK defines the primary input to the UK as coal, oil, gas, nuclear electricity and hydro-electricity. The latter inputs are converted to tons of 'coal equivalent according to the amount of coal needed to produce electricity at the efficiency of contemporary steam stations.'* This dubious procedure therefore introduces a theoretical loss into the energy supply system, as shown in Figure 3. The problem is thrown into focus by considering what energy cost should be attributed to one kilowatt-hour of electricity consumed. On the basis of the convention adopted by the *Digest of Energy Statistics* the total input is 22 784 x 10^6 therms, the output 5826 x 10^6 therms giving an efficiency of 25·57%. This corresponds to an energy cost of 3·91 kWhth per kWhe. However it could be argued that the electricity output of nuclear and hydro-stations is the true input to the system. On this basis the total input is 20 979 x 10^6 therms giving an efficiency of 27·77% so that the energy cost is 3·6 kWhth per kWhe consumed. Alternatively, it could be argued that the inputs to the system are the fossil fuels (19 903 x 10^6 therms), the heat generated at nuclear power stations (2360 x 10^6 therms) and the hydro-electricity output (194 x 10^6 therms).

This set of inputs gives an efficiency of 25·94% and an energy cost of 3·855 kWhth per kWhe. This last set of inputs is consistent

* A similar procedure is adopted for the electricity purchased from industry. The electricity is converted to 'tons of coal equivalent' and added to the 'coal input to the electricity supply industry'.

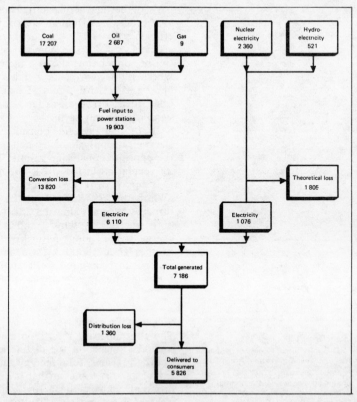

Figure 3. Energy network of electricity generation in the UK, 1968 (source: *UK Digest of Energy Statistics*). (units: 10^6 therms)

with a project examining the heat release[17] involved in the supply and consumption of electricity. If the project was investigating maximum possible efficiencies then the primary input to nuclear power stations is the energy theoretically available in the fabricated fuel rods. This may be an order of magnitude larger than the heat extracted from the rods in the reactor.

Apart from these differences there will also be alternatives in the ways in which the indirect energy consumption of power stations is taken into account. Indirect energy consumption is ignored in tables of energy statistics. In the case of nuclear power stations, how does one include the energy and materials consumed in processing and safeguarding nuclear wastes long after the power station has been taken out of commission? Table 1 summarises an analysis of the electricity supply industry based on the *Census of Production 1968*.[18] It includes the purchases of electricity from industry, the consumption of materials and equipment, the energy needed to refine oil, mine and transport coal etc. (It does not include estimates of the energy costs of nuclear wastes or fuel reprocessing.) On this basis the energy cost of 1kWhe is 4·19 kWhth, corresponding to an efficiency of 23·84%.*

In short this example shows that even when the sub-system is well defined there are alternative definitions of inputs which can dramatically alter the final values.

Oil refining

Another input to the energy supply industry is oil. Before any oil is consumed, either as fuel or chemical, it has to be extracted, transported and refined. The oil refinery is an easily identifiable sub-system; however there are differences both in what is counted as an input and how to partition the inputs between various outputs. A simplified flow diagram of the oil refinery system

[17] P.F. Chapman, *New Scientist* Vol 58, 1973 page 408.

[18] P.F. Chapman, 'The energy cost of delivered energy, UK 1968', ERG 003, Open University 1973

* In fact this value was obtained by solving the five simultaneous equations linking the major energy producing industries.

Table 1. The electricity supply industry

Inputs	10^6 kWhth
New buildings (£80·08 million)	1 895
Net plant etc (£462 million)	16 170
Net vehicles (£3·623 million)	178
Vehicle and machine spares	766
Iron and steel (13 500 tons)	134
Wire and cables	1 089
Other materials	985
Nuclear fuels	2 904
Coal (75·54 million tons)	542 015
Coke (314 000 tons)	2 944
Oil products	93 792
Gas	374
Electricity purchased (from industry)	3 479
Heat input (nuclear power stations)	96 113
Electricity (generated in hydro-plant)	3 600
Total input	766 438
Total electricity generated	215 149
Used in works, offices etc	16 195
Loss in distribution	16 182
Sold to final consumers	182 772
Overall efficiency =	23·84%
Hence 1 kWhe =	4·195 kWhth

connected to an 'organic chemicals industry' is shown in Figure 4. The problem is to decide how much energy to associate with an oil fuel and how much with an organic chemical. This is an interconnected system similar to that shown in Figure 2. This example has the added complication that the chemical feedstock is produced by the same plant as the oil fuel products.

Two of the many possible conventions that could be adopted in approaching this problem are that:

— since the crude oil is purchased as a primary fuel, all its calorific value is to be divided between the *fuel* products;
— the calorific value of the crude oil is to be divided between all the refinery products in the ratio of the calorific values of the products.

Following these rules the types of results obtained are:

- All the inputs are set as costs against the *fuel* outputs. The chemical feedstocks therefore have no energy cost attributed to them.* On this scheme the efficiency of the oil refinery as a fuel processor is 82·4%.[8] Thus the energy cost incurred in consuming a gallon of refined petrol is 53 kWhth, compared to an actual calorific output of 44 kWhth/gallon. Thus all industries consuming oil fuels are assigned greater energy costs than under the convention below; industries consuming chemical feedstocks have lower assigned energy costs.
- The sum of all the energy inputs is distributed as energy costs over all the refinery products in proportion to the actual calorific values of the various products. (Thus the energy cost of

* This is absurd because it indicates that 'energy costs' could be reduced by using a chemical as a fuel. This is a good reason for insisting that the 'energy cost' of a product should be greater than or equal to its calorific value. See also ref. 18.

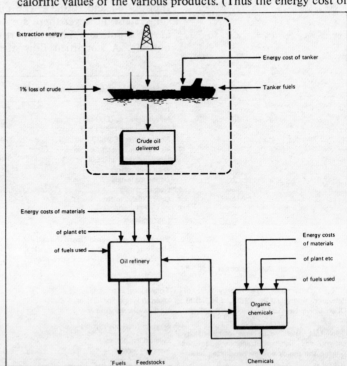

Figure 4. A simplified network of inputs and outputs associated with an oil refinery.

each product is its calorific value times a constant.) Note this increases the effective input to the oil refinery as compared to the convention above by virtue of the feedback of organic chemicals to the refinery. The efficiency of the refinery as a fuel processing plant is now 86%,[18] so the energy cost incurred in consuming a gallon of petrol is 51 kWhth. On this scheme 1 tonne of plastics has an average energy cost of about 30 000 kWhth; on the basis of the convention above the figure is between 10 000 and 15 000 kWhth.

The oil refinery is a sub-system which receives as an input the output of another sub-system, namely that shown as the extraction and delivery of crude oil in Figure 4. The energy costs incurred within the extraction and delivery sub-system are equivalent to about 7% of the calorific value of the crude oil.[8] However whether these energy costs are included as part of the input to the refinery depends upon the overall aim of the project. It would be consistent to ignore these energy costs in a project aimed at evaluating the energy costs of products *to the UK* since these costs are incurred outside the UK. If the project is associated with world energy costs then clearly these extraction and transport energy costs must be included as part of the input to the refinery.

Aluminium production

This final example of energy analyses shows how an understanding of different methods and conventions enables sense to be made of apparently contradictory results. Several independent estimates of the energy cost to produce 1 tonne of aluminium have been published. A statistical analysis of US data yielded 67 200 kWhth/ton;[4] an input-output analysis for the UK yielded 16 600 kWhth/ton[11] and process analyses have yielded 91 000 kWhth/ton,[9] 64 200 kWhth/ton[19] and 64 300 kWhth/ton.[20] The estimates made by US authors refer to the energy cost per *short ton*; converting these results to metric tons gives 75 000 kWhth/tonne by statistical analysis[4] and 71 900 kWhth/tonne[19] and 72 000 kWhth/tonne[20] by the process analysis methods. All these US studies assume a conversion efficiency for electricity of 33% whereas my own process analysis[9] assumes 23·8%. Converting my result to 33% electricity generation efficiency yields 72 000 kWhth/tonne, so that all three process analyses are in excellent agreement. The statistical analysis result is slightly higher; however it refers to the energy cost per ton of *rolled*, not crude aluminium.

The only remaining discrepancy is the value obtained by Wright[11] on the basis of an analysis of the UK input-output table. The 'aluminium' sector of this table does not distinguish between aluminium produced from primary ores and that produced from scrap material; nor does it specify whether the product is crude or semi-fabricated aluminium in the form of sheet, tube etc. Thus the energy cost deduced from the I/O table is an average energy given by

$$E_{av} = \frac{PE_p + SE_s}{P + S}$$

[19] J.C. Bravard, H.B. Flora, and C. Portal, 'Energy expenditures associated with the production and recycling of metals' Oak Ridge. Nat. Lab. Report. ORNL–NSF–EP–24, 1972.

[20] P.R. Atkins, *Engineering and Mining Journal* Vol 174, No 5, 1973, page 69

where P is the quantity produced from primary ores, S is the secondary production and E_p and E_s are the respective energy costs. Assuming that E_p equals 72 kWhth/kg, and E_s equals 3 kWhth/kg. This is close to the I/O table result; the remaining energy cost per ton of aluminium produced in the UK is about 14 kWhtht/kg. This is close to the I/O table result; the remaining difference could be due to some energy used to fabricate sheets etc.

Thus all these estimates of the energy cost of aluminium are self-consistent, they are all 'correct', but they are all based on different conventions. Clearly none of these results should be used in other studies until they have been converted to the conventions appropriate to other studies.

Conclusions

The analysis of energy consumption can show ways of conserving energy and can highlight particular types of problem. However this is a new area of study and there is no uniformity between authors as to what conventions to use or what techniques are most appropriate. This means that results from energy studies should be carefully interpreted and only used when the following points are clear:

- what sub-system of the world has been analysed;
- which energy inputs to the system have been included in the analysis;
- what calorific values are being used for primary fuels;
- what efficiencies are being ascribed to the energy supply industries;
- what conventions are being used to partition the energy costs within plants or industries.

Neglect of any of these factors could produce misleading conclusions even in work based on accurate data. Moreover, any work leaving room for doubt as to its conventions in respect of these points must be interpreted with great caution.

[21] P.F. Chapman, 'The energy costs of producing copper and aluminium from secondary sources', Open University, Report. ERG 002, 1973.

BIBLIOGRAPHY

A list of some of the important literature, with brief abstracts

Atkins, P.R., 'Recycling can cut energy demand dramatically,' *Engineering and Mining Journal*, May 1973, p 69
Energy cost of an aluminium can based on operating plant data. Based on direct energy consumption only.

Bravard, J.C. et al., 'Energy expenditures associated with the production and recycle of metals' ORNL-NSF-EP-24, Nov. 1972 (Available Oak Ridge National Laboratory, Oak Ridge, Tennessee 37830, USA)
Gives energy cost of magnesium, aluminium, iron, copper and titanium from ores. Does not count energy cost of transport or of machinery, water etc. Recycling energy cost is simply the energy to remelt the metal. Note all results refer to energy per short ton.

Boustead, I., 'Milk bottles and energy problems' *United Glass Magazine*, 1973.
Compares energy cost of glass and plastics milk bottles for delivering milk and indicates dependence on number of trips made. (Details of energy calculations available in TS251;CU9, 'Milk bottle', available from Open University, Milton Keynes, Bucks, UK)

Chapman, P.F. (a) 'The energy cost of producing copper and aluminium from primary sources', *Metals and Materials* Feb, 1974.
Gives complete energy costs based on process analysis and includes grade dependence of energy costs. (Details of energy calculations available in Research Report ERG001; Open University Energy Research Group, Open University, Milton Keynes, Bucks, UK)

(b) 'Energy conservation and recycling copper and aluminium' *Metals and Materials* (in press) 1974
Gives total energy cost of recycling including transportation and re-refining. Calculates potential energy cost saving which could result from increased recycling. (Details of calculations in Research Report ERG002, Open University.)

(c) 'The energy cost of delivered energy', UK 1968 Research report ERG 003, Open University.
Evaluates energy costs of coal, coke, gas, oil, and electricity from data in 1968 census of production.

(d) 'Energy and world modelling', Seminar report May 1973. (Available from OU Energy Research Group, Open University, Milton Keynes, Bucks)
Describes the relationship between energy cost and heat release and between heat release and climatic change. Indicates possible model based on energy flows.

Grimmer, D.P. and **Luszcynski, K.**, 'Lost Power', *Environment* Vol 14, No 3, 1972, p 14
Describes energy consumption of various transportation systems. Gives overall efficiencies of different systems but does not include energy costs of machinery, plant etc. Shows that electric automobile 20% efficient overall compared to gasoline auto efficiency of 10%.

Hannon, B.M., (a) 'Bottles, cans, energy', *Environment* Vol 14, No 2, 1972, p 11
Describes complete energy cost analysis of different containers with and without recycling.

(b) 'Aluminium cans', *Environment* Vol 14, No 6, 1972, p 46

Herendeen, R., 'The energy costs of goods and services', ORNL-NSF-EP-40 1972. (Availability: see Hirst)
Explains the input-output method of energy analysis developed at Oak Ridge and gives results for some services. (Details of results for US 1963 available in 'An energy input-output matrix for the US 1963' report, CAC 69; Centre for Advanced Computation, University of Illinois, Urbana, Illinois)

Hirst, E., (a) 'Energy consumption for transportation in the US' ORNL-NSF-EP-15, 1972. (Available from Oak Ridge, National Laboratory, Oak Ridge, Tennessee).
Energy costs devised from input-output analysis of US economy. Gives energy per ton-mile and energy per passenger-mile for different transport systems. (see also ORNL-NSF-EP-44, April 1973.

(b) 'Energy use for food in the US' ORNL-NSF-EP-57 (Available from Oak Ridge National Laboratory, Oak Ridge, Tennessee).
The energy use in food-related activities computed from input-output table for the year 1963. Shows total energy used by food industries is 12% of total energy. On average 6·4 Btu of primary energy consumed in delivery of 1 Btu of food energy.
This report is reviewed and brought up to date by E. Hirst, *Science*, Vol 184, 12 April 1974, p 134.

Hirst, E. and Herendeen, R. 'Total energy demand for automobile' *Society of Automotive Engineers*, Pamphlet 730065 (or 'How much overall energy does an automobile require', *SAE Journal of Automotive Engineering* Vol 80, No 7, 1972, p 36
Complete energy cost of automobiles based on input-output table analysis of US economy. Includes inputs from insurance, roads etc. Notable that fuel consumption only accounts for 60% of total energy cost.

Hirst, E. and Moyers, J.C., 'Efficiency of energy use in the US', *Science* Vol 179, No 4080, March 1973, p. 1229
Uses data on energy costs of transport (see Hirst and Hirst and Herendeen) and energy efficiency of space heating to outline strategy for energy conservation.

Leach, G., 'The energy costs of food production' in *The man-food equation* ed A. Bourne, Academic Press, 1973.
Wide-ranging review of energy in agriculture including energy cost analysis of major UK crops. Discusses implication of energy expensive food with respect to developing world and with respect to the price of energy.

Leach, G. and Slesser, M., 'Energy equivalent of network inputs to food producing processes' (Available: M. Slesser, Department of Pure and Applied Chemistry, University of Strathclyde, Glasgow, Scotland.)
Gives the energy costs of energy, fertilisers, transport etc based on published UK statistics.

Makhijani, A.B. and Lichtenberg, A.J., 'Energy and well-being', *Environment* Vol 14, No 5 (1972), p 10
Discusses relationship between total energy use and 'quality of life'. Gives complete breakdown of energy consumption in USA by 38 sectors. Includes energy cost of major materials. Energy costs deduced from annual statistics and are only approximate.

Mackillop, A., 'Low Energy Housing', *Ecologist* Vol 2, No 12, (1972)
Discusses energy cost of housing using energy costs of materials derived from 1968 Census of Production (UK). Energy costs are only direct fuel costs.

Mortimer, N., 'Energy cost of transport, UK 1968' (Research Report ERG004. Feb. 1973. Available from Open University Energy Research Group, Open University, Milton Keynes, Bucks)
Gives energy cost of road and rail transport in the UK which includes all indirect costs such as highways.

Pimentel, D. et al, 'Food production and the energy crisis' *Science* Vol 182, 2nd Nov 1973, p 443
Complete energy cost analysis of USA corn crops 1940-1970. Shows that while yields have increased dramatically the ratio of energy out/energy in has remained substantially the same.

Roberts, F., 'Energy consumption in the production of materials' *Metals and Materials*, Feb. 1974
Gives a review of the energy cost of major materials now and those to be expected in 30 years' time.

Roberts, P., 'Models of the future', *Omega* Vol 5, No 5, 1973, p 592
Discusses the role of energy in an industrial system and in models of the future. Indicates the significance of energy cost in imposing constraints, especially the energy cost of energy.

Slesser, M., 'Energy subsidy as a criterion in food policy planning', *Journal of the Science of Food and Agriculture*, Vol 24, Nov. 1973
Review of energy costs of a wide range of crops showing that they conform to a trend relating to yield and energy output.

Smith, H., 'The cumulative energy requirements of some final products of the chemical industries' *Transactions* (World Power Conf.) Vol 18 1969
Gives energy cost of a number of important chemicals and shows that the analysis of chemical industry is complex. Energy cost does not include energy in feedstocks nor all energy costs of plant, vehicles etc.

Wright, D.J., 'Calculating energy requirements of commodities from the input/output table' (Available: Systems Analysis Research Unit, Dept. of the Environment, Marsham Street, London)
Outlines basis of method and gives results based on UK input-output table (70 sectors).

2. The energy cost of fuels

Peter F. Chapman, Gerald Leach and M. Slesser

In the UK, the energy industries are the largest *consumers* of energy, using more than 30% of the total primary energy input. The first part of this chapter examines the five energy industries – coal mining, oil refining, coke, gas and electricity production – using the data provided in the *Report on the UK Census of Production 1968*. The results obtained are compared with those from other sources, and in the second part of the chapter an attempt is made to put these results in a wider context by examining some of the implications for alternative sources of energy.

The events of late 1973 made it transparently obvious that energy, in the form of fuel supplies, is crucial to a modern industrialised society. This rude awakening to the significance of energy has led to increased research into future sources of supply and, very belatedly, research into how energy is utilised in a modern economy. At last we are asking ourselves the question 'how much energy do we *need*?' This question has no definitive answer; any answer will depend on the assumption made about standards of living, personal mobility, sources of food and materials and, of course, possible technological developments. One way of approaching this question is to examine how we presently make use of the energy consumed.

Perhaps the most startling result of this examination is that the largest energy-consuming sector of our economy consists of the energy industries. The five energy, or fuel, industries – coal mining, oil refining, coke, gas and electricity production – jointly consume more than 30% of the total energy input to the UK. Put another way, for every 100 units of energy or fuel input to the UK less than 70 units is delivered to a consumer for use.

There are three reasons for examining the energy efficiency of the fuel industries in more detail:

- the fact that the fuel industries are themselves the largest consuming sector offers the chance of reducing the demand for primary energy supply without adversely affecting the rest of the industrial system;
- the energy wasted by industries constitutes a major hazard to local,[1] and perhaps global,[2] climate. Whatever the limit on the 'safe' heat release into the atmosphere it is clearly desirable to minimise the ratio of the 'heat wasted' to energy delivered to consumers;
- it is essential to know the energy efficiency of individual fuel industries in order to evaluate the energy costs of manufactured products or processes.[3] With this information it is possible to compare the total efficiency of two processes which consume, say, one ton of coal or 1000 kWh of electricity.

[1] C.I. Griffiths, *Meteorological Magazine*, March 1974

[2] For example, see 'Inadvertent climate modification' SMIC (MIT Press, USA, 1971), or P.F. Chapman, *New Scientist*, Vol 47, 1970, p.634

[3] Chapter 1: 'Energy costs: a review of methods'.

The fuel supply industries form a complex interconnected system with each industry supplying fuel to every other. For example, oil refineries supply fuel to electricity generating stations which supply electricity to the oil refinery. An oil refinery may also provide the fuel used by the tankers which deliver crude oil to the refinery. These interactions require a careful method of analysis based on a systems approach. In addition to consuming fuel the fuel industries also consume large quantities of materials and machines which require energy for their production. This 'indirect' energy consumption must be included in the analysis. Finally, most of the industries produce more than one product as output. It is therefore essential to have an acceptable convention for partitioning the total energy costs of the inputs between the different outputs.

The first part of this chapter outlines the method used to analyse the energy supply industries using the data provided in the *Report on the UK Census of Production 1968*.[4] The results obtained are compared with those derived from other sources. The second part of the chapter attempts to put these results in a wider context by examining some of the implications of alternative sources of energy.

Method of analysis

The maximum energy which can be extracted from a fuel is called the *calorific value* of the fuel. For example, the heat energy available in one ton of coal is about 8000 kWh, the exact value depending upon the type of coal. When the coal is burned some of the available energy may not be utilised either because the coal is only partially combusted or because some of the heat generated is 'lost' up the chimney. The calorific value of coal is therefore a measure of the amount of heat energy potentially available.

In evaluating the energy cost of a product it is the total energy available which is counted as part of the energy cost, not simply that part of the energy which is utilised. However, it is not sufficient to consider simply the calorific values of the fuels used in a particular process; account must also be taken of the energy expended in making the fuel available for use. For example, the mining and transport of coal involve the consumption of fuel, so the total energy cost associated with the consumption of a ton of coal is the sum of its calorific value and the energy expended in producing the ton of coal. The sum is called the energy cost of coal. It is the purpose of the present analysis to evaluate the energy costs of fuels as delivered to industrial or domestic consumers.

The network of industries in the UK can be divided into two sectors, as shown in Figure 1. The first sector comprises the five fuel industries and has, as basic inputs, the raw fuels: oil in the ground, coal in the ground etc. Other inputs are machinery, plant, equipment, materials and services, such as transportation, from the industrial sector. The industrial sector is, in this model, a large 'black box', which consumes fuels and raw materials to produce final commodities. The goal of energy analysis is to apportion the total energy input in the form of primary fuels between

[4] *Report on the Census of Production, 1968* (HMSO, UK, 1971)

The energy cost of fuels

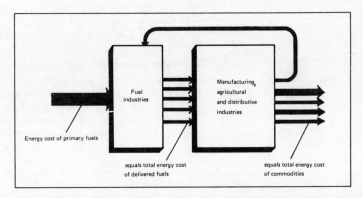

Figure 1. The 'fuel supply' and industrial sectors

commodities. Thus for the total system there is a convention of 'energy cost conservation'. This states that the sum of all the inputs (x_i) times their respective energy costs (E_i) should equal the sum of all the outputs (y_j) times their respective energy costs (E_j). Formally this can be written:

$$\underset{\text{inputs}}{\Sigma \, x_i \, E_i} = \underset{\text{outputs}}{\Sigma \, y_j \, E_j}$$

When the analysis is complete the energy costs of the outputs (E_j) reflect the proportion of the primary fuel inputs required for their production.

The present analysis is concerned with the inputs and outputs of the fuel industries. It is important to note that the conservation of energy cost is *not* the same as the conservation of energy since, for example, the energy flowing into the fuel industries is *not* equal to the energy content of the fuels flowing out.

The steps in proceeding with this analysis are first to identify all the inputs and assign energy costs to each and second to identify the outputs and calculate their energy costs.

The fossil fuel inputs to the fuel industries — oil, gas and coal — are given energy costs equal to their calorific values. There are two other primary inputs, namely nuclear fuels and hydro-electricity. There is no generally acknowledged 'calorific value' for a nuclear fuel and various authors count this input in different ways. In this analysis the nuclear input is given an energy cost equal to the heat generated in the nuclear reactor. This is compatible with using the calorific values of fuels since, in principle, the heat from a nuclear reactor could be substituted for the heat obtained by burning coal. For hydro-electricity the energy cost is taken as the electricity output since this is the *heat* equivalent. (In practice hydro-electric installations are between 80% and 90% efficient at converting mechanical energy into electricity.)

The remaining inputs to the fuel industries are materials, machines and transport, all products of the industrial sector. The energy costs attributed to these items are those deduced by a preliminary analysis of the 1968 Census Report. These inputs represent about 2·5% of the total energy cost input,[5] so the use of approximate energy costs (accurate to ± 10%) for these does not involve significant errors in the final result.

[5] P.F. Chapman, 'The energy cost of fuels: UK 1968', Open University Energy Group: Research Report ERG003, November, 1973

The energy cost of fuels

The outputs of the fuel industries involve a different type of problem, namely deciding how to apportion the input energy costs between the different outputs. There are four obvious possibilities.

1. Assign all the energy costs to the principal output, since this is required product.
2. Assign the energy costs on a financial basis so that every output of an industry has the same energy cost per £ value.
3. Assign the energy costs on a weight basis so that every output has the same energy cost per ton.
4. Assign the energy costs on the basis of calorific value so that all outputs have the same ratio of energy cost to calorific value.

Assumption 1 implies notions of purpose or usefulness inappropriate to a study based on physical variables. It also results in logical absurdities when the same end product (gas) is the principal product of one plant (gasification plant) but a secondary product of other plants (coke ovens etc). Assumption 2 is convenient since then the flow of energy can, in principle, be traced through a system by techniques of financial analysis (eg the input-output table method).[6] It is a dangerous procedure since the price of products changes with time and may be different to different purchasers. The partitioning on a weight basis (Assumption 3) is an attempt to relate the energy cost to a physical property of the product. However, this and all the other procedures outlined above could result in the absurd situation of having the energy cost of a product less than its true calorific value. (This is absurd because the analysis could then make recommendations for saving 'energy costs' without altering the real consumption of energy!) Thus the only convention which is physically sensible is to apportion energy costs on the basis of calorific value. This is the basis of the method described below.

If the analysis is to be based on calorific values then there are two sets of data required. The first is the physical quantity (in tons, gallons etc) of each output of a given industry. The second is the calorific value of each of the outputs. This procedure allows an independent check on the data for each industry since no industrial plant should produce more calorific value than it consumes.

Having identified all the inputs (x_i) and assigned energy costs (E_i) to them and identified all the outputs (y_j) and their respective calorific values (C_j) it is now possible to evaluate an 'efficiency' for each industry. This efficiency is defined as being equal to the total calorific value of outputs divided by the total energy cost of inputs. It is thus a *conventional* efficiency and *not* a true efficiency like energy out divided by energy in. Denoting the efficiency by η this can be written

$$\eta = \frac{\text{calorific value out}}{\text{energy cost in}}$$

$$= \frac{\Sigma \, y_j \, C_j}{\Sigma \, x_i \, E_i}$$

[6] See Chapters 6-8.

Thus the energy cost required to produce one unit of calorific value output is the reciprocal of this efficiency. The reciprocal of 'efficiency' can be denoted by ϵ, so that

$$\epsilon = \frac{1}{\eta}$$

The energy cost of an output is the proportion of the energy cost of the inputs required for its production. The input energy cost required for each unit of calorific value output is simply ϵ. Thus if an output has a calorific value C_j kWh/ton then its energy cost will be $\epsilon\, C_j$ kWhth/ton:

$$E_j = \epsilon\, C_j$$

Thus, provided the constants, ϵ, appropriate to each industry can be evaluated, the energy cost of an output can be deduced from its calorific value. Herein lies a problem, for each energy industry actually supplies fuels to each of the other fuel industries. Thus to calculate the efficiency (or ϵ) for the electricity industry we first need the efficiencies of all the other fuel industries, since these are inputs to the electricity industry. But electricity is also an input to all the other industries!

This problem is solved by setting up the five simultaneous equations describing the inputs and outputs of each industry. The procedure is as follows:

(a) Identify all the outputs for industry A, find their calorific values and calculate the total calorific value out, C_a.
(b) As explained above, the total energy cost of the output of this industry is then equal to $\epsilon_a C_a$.
(c) Identify all the inputs to industry A other than those from other fuel industries. Assign energy costs to each input and work out the energy cost input from these sources, E_a.
(d) Calculate the total calorific value of the input to industry A from another fuel industry, say industry J. Call this C_{ja}. The energy cost of this input is equal to $\epsilon_j C_{ja}$ where ϵ_j has not yet been found. This is repeated for all the other energy industries which supply industry A.

Now we can use the 'conservation of energy cost' to obtain the equation for industry A:

energy cost out = Σ energy cost in

$$\epsilon_a C_a = E_a + \sum_j \epsilon_j C_{ja}$$

This equation has only five unknowns (the five values of ϵ for each industry) and we can obtain five such equations. So the equations can be solved and the values of all the ϵ's found.

Solving the equations

Applying the technique outlined above to the five UK fuel industries results in the five energy cost equations[5] which summarise the operation and interdependence of the fuel industries in 1968 (see Table 1). The multipliers, ϵ_m etc, are the reciprocals of the efficiencies of the industries involved. Thus 'reading' the first equation from left to right gives:

Table 1. Equations linking the energy industries (1968)

$1160.94\epsilon_m$	=	1187.63	+		+	$0.146\epsilon_c$	+	$0.04\ \epsilon_g$	+	$0.24\epsilon_o$	+	$5.07\epsilon_e$
$223.0\ \epsilon_c$	=	0.41	+	$235.45\epsilon_m$	+		+	$10.60\ \epsilon_g$	+	$1.14\epsilon_o$	+	$0.41\epsilon_e$
$187.74\epsilon_g$	=	7.72	+	$81.09\epsilon_m$	+	$66.6\ \epsilon_c$	+		+	$75.56\epsilon_o$	+	$1.09\epsilon_e$
$1077.5\ \epsilon_o$	=	1092.2	+	$0.04\epsilon_m$	+	$75.73\ \epsilon_c$	+	$25.25\ \epsilon_g$	+		+	$1.34\epsilon_e$
$182.77\epsilon_e$	=	127.3	+	$520.42\epsilon_m$	+	$2.48\ \epsilon_c$	+	$0.269\epsilon_g$	+	$82.27\epsilon_o$		

Note: m = coal mining; c = coke; g = gas; o = oil; e = electricity. All numbers are $\times 10^9$ kWh

- calorific value of coal mining output is 1160.94×10^9 kWh
- this calorific value times ϵ_m is the energy cost out
- the energy cost out *equals* the sum of the terms on the right hand side which are energy cost inputs
- the energy cost input from materials consumed and coal in the ground is 1187.63×10^9 kWh.
- the energy cost of fuel purchased from the coke industry is ϵ_c times 0.146×10^9 kWh
- the energy cost of fuel purchased from the gas industry is ϵ_g times 0.04×10^9 kWh etc.

The total energy cost input to all the fuel industries from primary fuels and raw materials is the sum of all the terms on the right hand sides which are *not* multiplied by an ϵ_j. Thus the primary input to the energy industries is $1187.63 + 0.41 + \ldots = 2415.26 \times 10^9$ kWh. The *net* energy output of the energy industries has to be calculated by subtracting the fuels delivered to other energy industries from the gross output. Thus the gross calorific output of the coal industry is shown as 1160.94×10^9 kWh. But of this 235.45×10^9 kWh was delivered to the coke industry (shown in second equation), a further 81.09×10^9 kWh to gas, 0.04×10^9 kWh to oil and 520.42×10^9 kWh to electricity. The *net* output of the coal industry is thus 323.94×10^9 kWh. The *net* output of all the fuel industries is 1646.7×10^9 kWh. Thus the *overall* efficiency of the fuel sector is 68.17%, indicating a loss of more than 30% of the primary energy input (see Figure 2).

Solving these five equations for the five unknowns, ϵ_c, ϵ_m etc., gives the values and corresponding efficiencies set out in Table 2. Also shown in Table 2 are similar results deduced from the 1963 Census Report[7] and values deduced from less detailed data in the

[7] T. Jackson and P.F. Chapman, 'Analysis of the 1963 Census of Production' (to be published)

Table 2. Efficiencies of energy industries

	1968		1963	1971/72
Industry	(from Table 1) (ϵ)	Efficiency (%) ($1/\epsilon \times 100$)	Efficiency (%)	Efficiency (%)
Coal	1.042	95.99	95.49	95.5
Coke (ϵ_c)	1.181	84.71	75.54	88.0*
Gas (ϵ_g)	1.390	71.92	64.74	81.1
Oil (ϵ_o)	1.134	88.21	80.82	89.6
Electricity (ϵ_c)	4.192	23.85	22.02	25.2

* This result is less accurate than others due to lack of data for coke-ovens 1971/72

The energy cost of fuels

Figure 2. A summary of the energy flows of the energy sector as a whole. Quantities in 10^9 kWh

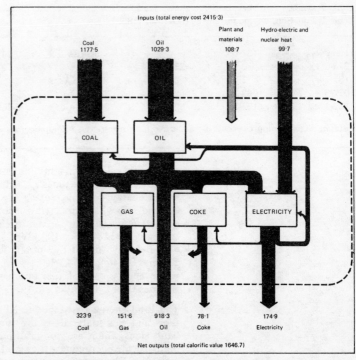

[8] *United Kingdom Energy Statistics, 1973* (HMSO). Note this publication does not give full details of all purchases and the final results are less accurate than those based on refs 5 and 7

[9] D.P. Grimmer and K. Luszczynski, *Environment*, Vol 14 No 3, 1972, p 14

[10] G. Leach, *The motor car and natural resources* (OECD, Paris 1972); and G. Leach, 'The impact of the motor car on oil reserves, *Energy Policy*, Vol 1, No 2, 1973, p 195

[11] *Electricity supply statistics*, (Electricity Council, London, 1971)

UK Energy Statistics[8] for 1972. There is a number of important factors to bear in mind in interpreting these results.

Perhaps the most important is that the efficiency of a fuel industry should not be considered in isolation but in conjunction with typical efficiencies for utilising that energy. For example, a careful comparison of the overall energy efficiencies of electric[9] and petrol[10] powered cars, summarised in Figure 3, shows electric cars to be only marginally less efficient. A similar comparison between oil-fired and electric-powered house heating (Figure 4) shows the oil system significantly more efficient. Thus the efficiency of the supply industry is only part of the overall 'energy efficiency'.

The second point to note is that the significant improvement in the efficiency of the gas industry in 1972 is, to a large degree, due to the use of natural gas as opposed to town gas. Although this probably reflects a true gain in efficiency the data available do not include all the exploration and drilling 'energy costs' of providing natural gas.

The overall efficiency of the electricity supply industry is much lower than the notional 33% assumed by many authors, presumably on the basis of modern power station efficiencies of 35%. There are three reasons for this. First, *most* of the power stations operating in the UK are not 'modern' and the overall thermal efficiency is still less than 29%.[11] Figure 5 shows the steady, but slow, increase in 'overall thermal efficiency' achieved since 1932. Second, the transmission of electricity involves losses. In 1968 these amounted to 7·5% of the electricity generated.[4] A further 7·6% of the electricity generated[4]

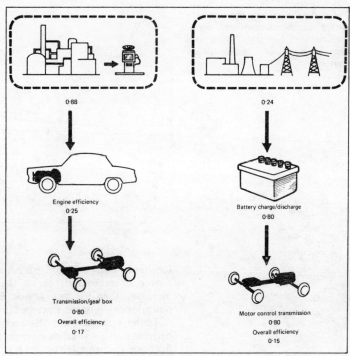

Figure 3. Comparison of the total efficiency of electric and petrol powered cars

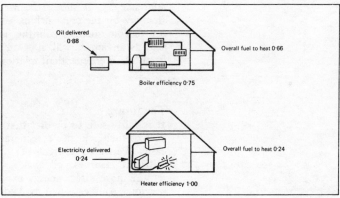

Figure 4. Comparison of the total efficiency of electric and oil-fired house heating

was used in electricity offices, works, showrooms etc. The third reason why the overall efficiency is lower than normally assumed is that this analysis has incorporated the energy costs of materials, plant, equipment etc, consumed by the industry, as well as the true energy costs of the fuels delivered to power stations.

These results in no way reflect the 'technical' efficiencies of the respective industries in the sense of showing what fraction of the potential energy is actually delivered to final consumers. For example, no account is taken of the oil left in a well when it ceases to be worked, nor of the coal not recovered from a deposit. Similarly, no account is taken of the theoretical energy available in a nuclear fuel rod, an energy which may be 100 times larger than that recovered in a burner reactor. To some degree it seems

The energy cost of fuels

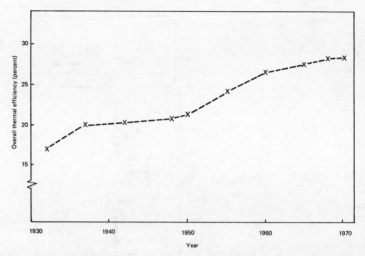

Figure 5. The overall thermal efficiency of the conventional steam stations in the UK. (Note this excludes hydro and nuclear power stations.)

inevitable that as more of the energy potentially available in a source is recovered, ie as the 'technical recovery efficiency' increases, then the 'energy cost efficiency' will decrease. This is a point taken up in the next section.

Table 3 summarises the energy costs of the fuel products as delivered to final consumers in the UK in 1968. These data are fundamental inputs to energy cost analysis, (as explained in reference 3). It should be mentioned that these results depend crucially on the conventions set out previously. Changing any one convention, such as counting nuclear electricity and not nuclear heat as an input, will alter all the values since they are derived from the interdependent relationships set out in Table 1.

The future

It is fairly safe to predict that over the next 20 years the energy cost of fuels will rise considerably because of three factors. First, easily recovered sources of fuel are steadily being replaced by 'difficult' sources: oil shales, North Sea oil and nuclear power require more direct energy expenditure for their production than previous sources such as Middle East oil. Second, many new energy technologies involve conversions from primary fuels to secondary fuels such as gasification or liquefaction of coal, and production of hydrogen electrolytically or chemically. The third factor is that as the rich sources of materials are exhausted, the energy cost of materials will rise considerably. Since the new fuel sources also require more materials input per unit output this factor may significantly increase the energy cost of the fuel production. Together these factors are inflationary. A rise in the energy cost of fuel will increase the energy cost of materials (since material production consumes fuel). The increase in energy cost of the materials is further increased by the lower grade of ore. In its turn this increased energy cost of materials increases the energy cost of the fuel production process – hence energy cost inflation. To date this inflationary effect has been counteracted by

Table 3. Energy costs of products of the fuel industries

	1968		1971/72	
	(kWhth/ton)		(kWhth/ton)	
Coal				
to iron and steel	8 056		8 100	
to chemicals	7 600		7 640	
to china and glass	8 608		8 650	
to cement	7 509		7 550	
industrial average	8 334		8 380	
Coke	9 340		8 990	
coke breeze	7 610		7 320	
other solid fuels	9 340		8 990	
Oil products	15 013	(49·9/gall)	14 780	(49·1/gall)
Motor spirit	14 547	(54·5/gall)	14 330	(53·6/gall)
Derv (diesel fuel)	13 718	(58·4/gall)	13 510	(57·5/gall)
Fuel oil Chemical feedstock	15 279		15 050	
Gas		40·73/therm		36·1/therm
Electricity		4·192/kWh		3·97/kWh

improvements in technical efficiency, but there is little room left for further improvements.

This energy cost inflation of obtaining future fuel supplies has serious policy implications so it is worth examining the basis of the argument in more detail.

The mining industry has developed impressive mechanised techniques for winning materials. This has been accompanied by decreasing financial costs but rising energy costs. In common with other industries financial savings have been made by decreasing labour costs using energy intensive technologies. For example, over the past 50 years the annual output tonnage of all US mines has increased by about 50% whereas the annual fuel consumption has increased by 600% in the past 25 years.[12] Figure 6 shows the expected increase in energy cost per kilogram of copper as the grade of ore decreases.[13] This type of variation will occur for all the relatively scarce metals such as zinc, lead, nickel, tin etc. For the relatively abundant metals, notably iron and aluminium, the increases are expected to be considerably less dramatic, but still significant.[14] The energy costs of plastics and petrochemicals are clearly directly tied to the energy cost of fuel sources (see below). The remaining materials used in significant quantities: glass, cement, bricks, etc, require substantial quantities of fuel for their production but are not subject to scarcity of raw materials.

Thus there is a trend towards greater energy costs associated with the production of materials from primary sources. The only obvious way of off-setting these increases is to increase the proportion of materials supplied by recycling. For metals the

[12] T.S. Lovering in *Resources and Man*, (W.H. Freeman & Co, San Francisco, 1969), p 122
[13] P.F. Chapman, *Metals and Materials*, February 1974, p 107
[14] J.C. Bravard, et al, *Energy expenditures associated with the production and recycle of metals*, (ORNL-NSF-EP-24 Oak Ridge Tenn., USA, 1972)

The energy cost of fuels

Figure 6. The energy cost per kg of copper as a function of ore grade. The solid curve is based on optimistic assumptions, the dashed curve on current technology

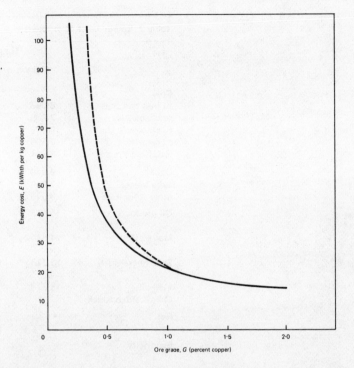

energy cost of recycling is generally an order of magnitude less than production from ores.[15]

While growth in total material consumption continues there are very stringent limits on the proportion of consumption that can be met by recycling. (This arises because the recycled material is a fraction of consumption some time in the past, which, under growth conditions, is less than present consumption.) So unless the rate of growth is *reduced* and the recovery of materials from scrap *increased*, the energy costs of materials will increase in the future.

There are presently a number of studies underway to try to evaluate the likely energy costs of future fuel supplies. To date only approximate data are available and these must be treated with caution. However, the trend is significant. The energy required to extract one ton of crude oil from a Middle East well has been estimated to be 500 kWhth/ton,[5] representing about 4% of the fuel energy obtained. Transportation to the UK absorbs a further 5% of the fuel[16], 4% as tanker fuel and 1% as loss in ballast. In comparison, the energy cost of an oil rig suitable for the North Sea represents about 10% of its total fuel output.[17] Assuming that the extraction costs represent a similar fraction as for the Middle East and that pumping ashore requires a further 4% of the fuel energy the total energy cost rises to about 18% of the fuel output. Thus an oil industry based on North Sea oil may have an efficiency as low as 80%, compared with the present 88%. And this estimate has not included energy costs associated with exploration, an activity involving considerably more material and fuels in the North Sea than in the Middle East.

[15] P.F. Chapman, *Metals and Materials*, June 1974

[16] G. Leach and M. Slesser, *Energy equivalents of network inputs to food producing processes.* (University of Strathclyde, Scotland, 1973)

[17] Based on cost estimate of rig and kWhth/£ value deduced from analysis of Census of Production. Pumping energy is large because secondary recovery process used from the start

The energy cost of fuels

In a similar way the net efficiency of the electricity industry may decrease as the proportion of nuclear power increases. The efficiency of future nuclear power stations may rise to 36%, the best efficiency so far achieved in a conventional station. However, the nuclear fuels used in burner reactors are very energy expensive. About 5% of the output of the power station is required to operate the gaseous diffusion plant for enriching uranium.[18] To mine the 0·7% U_3O_8 ores currently being developed will require a further 1-2% of the station output. Presumably, 15% of the output will be consumed as at present, in power stations, offices, showrooms and distribution losses. Furthermore, the energy cost of constructing the power station and appropriate proportions of the fuel preparation and reprocessing plants is equivalent to 1½ years' output of the power station. Assuming a 25-year lifetime, this represents a further loss of output equivalent to 6%. All these energy expenditures reduce the 36% station efficiency to an overall efficiency of 25·9% as shown in Figure 7. This, however, is not the end of the story. No account has been taken of the energy costs associated with research and development and, more importantly, those associated with waste disposal and protection. These are difficult items to incorporate in such an analysis, partly because suitable schemes have yet to be developed. (An energy cost point of view poses serious questions for proposals such as shooting wastes into the sun!) However, it seems likely that any suitable scheme will require husbanding these wastes for many, many years after the station has ceased operating — implying a continuing energy expenditure even after the source has ceased to supply energy.

Preliminary examinations of processes for extracting oil from shales or tar sands show these to be particularly energy expensive processes.[19] In summary, these current developments for obtaining

[18] N. Mortimer, (private communication based on approximate analysis of nuclear reactors)

[19] This is because in both cases significant quantities of rock have to be mined, crushed and processed to extract the oil. This involves considerably larger energy costs than drilling holes and pumping the oil up.

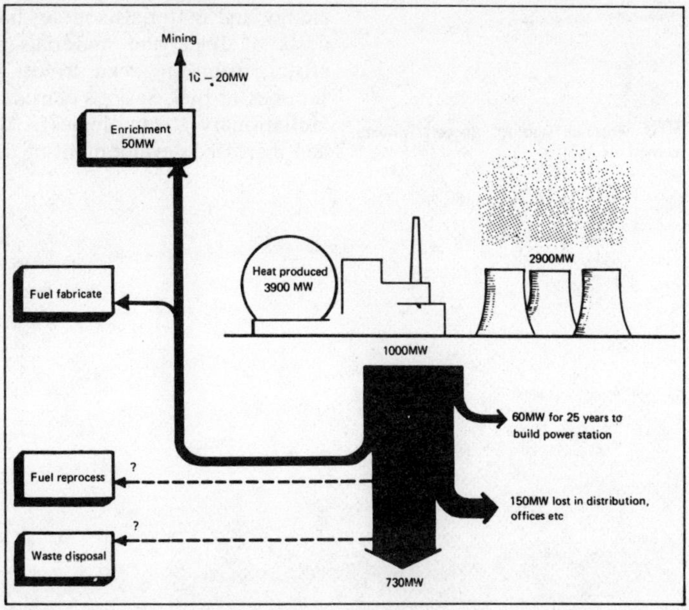

Figure 7. Approximate energy flows associated with a 1000 MW nuclear power station

increases in fuel supply do not offer the likelihood of reducing the energy cost of fuel; instead the energy costs may rise.

One serious consequence of any such rise is that it brings much closer the time when we have to concern ourselves with the climatic effects of heat release.[2] Virtually all the energy consumed ends up as heat in the atmosphere, only a small fraction being converted to (fixed) chemical energy. Thus if the energy costs of fuel and materials increase, the heat release associated with a given standard of material living will also increase. This will be further increased by any trend towards electricity as a major power source (since it has the lowest efficiency).

The climatic effects of heat release could be avoided by the development of technologies able to exploit the income, or renewable, energy resources. The use of wind, hydro, solar or geothermal power does not constitute an additional heat input to the atmosphere and could affect local climate only if energy were generated in one region and transported to another. It remains to be seen whether these income sources can be exploited for low energy costs. Preliminary calculations on the energy costs of solar cells[20] shows that the cell has to operate for about 10 years before the energy of fabrication is recovered. In contrast, the energy cost of windmills appears to be such that the energy of production is recovered in two or three years.

Conclusion

The energy cost of fuels should be an important consideration in formulating future energy policies. The analysis presented here has shown that at present the energy sector of the UK economy consumes more than 30% of the primary energy input. The discussion of future supplies of fuel and materials indicates that this proportion may be increased substantially as lower grade energy and materials sources have to be used. Together the energy costs of fuels and materials produce an inflationary tendency arising from the need to use more energy to obtain additional supplies of fuel. Serious consideration should therefore be given to 'deflationary' technologies such as increased materials recycling and increased development of the use of renewable energy sources.

[20] G. Turnbull and M. Slesser, (private communication)

3. The energy costs of materials

Peter F. Chapman

Next to the energy industries, the materials industries are the most energy intensive in the UK. This chapter analyses energy costs using the conventions described in Chapters 1 and 2 for metals, building materials and paper; polymers are the subject of Chapter 4. Four ways of saving fuel in materials production and processing are identified and discussed: improvement of fuel efficiency; substitution of less energy intensive materials for more energy intensive ones; greater recycling of materials; reduction in use of materials.

A modern industrial society has been described as resting upon the tripod of materials, energy and information. All aspects of our culture involve a mix of these three basic ingredients. But they are not independent. The communication of information requires energy, energy conversion requires the use of materials and the extraction and production of useful engineering materials requires energy. This chapter concentrates on this last interaction, the energy required to produce materials. There are a number of reasons why that is of interest.

Firstly, the materials production industries comprise the largest fuel consuming sector of the economy apart from the fuel industries themselves. As will be shown later materials production accounts for almost 30% of delivered energy consumption in the UK and worldwide accounts for over 20% of world fuel consumption. This indicates that savings in fuel consumption in these industries could have a substantial impact on total fuel demand.

Secondly, the materials industries are, together with the fuel industries, the most energy intensive industries. The energy intensity of an industry is determined by its ratio of direct fuel consumption to value added. In the UK the materials industries contribute 7%[1] to the GNP but consume 30% of all fuels indicating that they are four times more energy intensive than the UK average. This means that these industries are those most vulnerable to fuel price rises.

Thirdly, although the materials industries are very energy intensive the fuel purchases only represent 10 to 15%[1] of the total cost of producing materials. This means that the relative energy costs of materials may not be reflected in their relative financial costs. Consequently engineers and designers may unnecessarily increase total fuel consumption by choosing a material of higher energy cost (but lower or similar financial cost) than necessary. If energy conservation is to become part of good design then the energy costs of materials are necessary data.

Finally the methods of analysis can be extended so as to deduce the energy costs of producing materials from different sources or sources of different grades. This enables realistic estimates of the future costs of materials to be made. When combined with

[1] Deduced from the data presented for the materials industries in *Report on the Census of Production, 1968*, HMSO, 1971.

information on the future demand for materials these results provide an important input for estimating future energy demand and hence the determination of energy policy.

Energy costs

The difficulties of interpreting energy costs deduced using different methods and conventions employed by various authors have been described in Chapter 1. Most of the results presented in this paper have been obtained by analysing the data presented in the 1968 Census Report.[1] The conventions used in this analysis were described in Chapter 2; briefly they are:

- the inputs to an industry which are given an energy cost are purchased fuel, materials, capital equipment and transport
- the energy costs ascribed to the inputs are themselves deduced from the Census Report
- the energy costs of fuels are those described in Chapter 2.
- the energy conversion efficiency for generating electricity is 23.85%
- the energy cost of chemical feedstocks includes the calorific value of the feedstock

Table 1 gives the energy costs of the major materials produced in the UK deduced from the 1968 Census. Also shown are two energy coefficients. The first refers to the energy intensity of the industry and equals the direct energy consumed divided by the

Table 1. Energy costs of materials produced in the UK, 1968

	Energy cost (kWhth/ton)	Direct energy cost £ value added (kWhth/£)	Total energy cost £ value output (kWhth/£)
Metals			
Crude steel	10 540		
Finished steel	13 200	402	212
Aluminium	27 000	87	74
Copper	12 750	60	32
Zinc	19 000	111	45
Lead	7 000		
Building materials			
Glass	6 250	163	142
Wood	1 800	21	
Cement	2 200	802	410
Plaster	900		
Plaster board (per sq yd)	15	93	144
Roof tiles (per sq yd)	43		
Bricks	500	253	173
Other			
Paper	7 500	247	191
Plastics	45 000	150	213
Synthetic rubber	41 000		
Chemicals (organic)	—	832	455
Chemicals (inorganic)	—	275	194

value added by that industry. The second coefficient equals the energy cost of fuels consumed divided by the total value of its output. This second coefficient can be used to estimate the relative price changes of the materials with changes in fuel prices.

Although the convention described above (see also Chapter 2) provides a consistent basis for the analysis of fuel consumption in the UK it is inadequate when discussing world-wide fuel consumption. The UK results represent the particular mix of technologies appropriate to the UK sources of materials, in many cases these are not typical. This can best be illustrated by comparing the above results with those of other authors.

Steel

There are several factors which make it difficult to deduce an unambiguous energy cost for steel. Firstly different workers quote values for crude steel while others give values for finished steel. Furthermore the quantity of crude steel required to produce one ton of finished steel varies between 1·2 and 1·5 tons of crude steel. Secondly the iron and steel industries are sometimes aggregated in published statistics; sometimes only data for steel companies is given. Thirdly the iron and steel sector sells energy, in the form of electricity and gas, to other sectors. These sales have to be deducted from the purchased energy to deduce a net energy cost. Fourthly it should be emphasised that the output of a steel industry is not homogeneous. A semi-finished slab of steel may have an energy cost as low as 8000 kWhth/ton whereas a thin sheet of special alloy steel which has been specially heat-treated may have an energy cost as high as 60 000 kWhth/ton. Finally there are different conventions for assigning energy costs to the steel scrap consumed in a modern steelworks. The value given in Table 1 assumes that the scrap purchased by the steel works has an energy cost equal to the purchase price times the energy coefficient for transport (45 kWhth/£).

Provided the same energy costs of fuels are used the energy cost of steel given in Table 1 is within 5% of the value calculated from the Iron and Steel Annual Statistics.[2] A number of American studies of the energy cost of steel give the following results (when converted to metric tons): 14 112 kWhth/ton;[3] 16 654 kWhth/ton[4] and 15 613 kWhth/ton.[5] All these authors use an electricity conversion factor of 33% which should make their results lower than that in Table 1 by about 4%. In fact the average of these US values is 17% higher than the UK result suggesting that the UK industry is between 15 and 20% more efficient than the US industry. In view of the uncertainties listed above all these results are in good agreement.

Other metals

For the remaining metal industries, copper, aluminium, zinc, lead etc. the UK industries are not typical of world-wide production. In the UK a significant proportion of all these metals is imported and another, large, fraction recovered from scrap metal. In contrast most metal in the world is produced from primary ores. The UK average energy costs for these metals are thus considerably below the world average. Table 2 gives the best values available for the

[2] *Iron and Steel Industry: Annual Statistics for the United Kingdom, 1972* (and previous years), British Steel Corporation (for Iron and Steel Statistics Bureau, P.O. Box No 230, 12-16 Addiscombe Road, Croydon, CR9 6BS, UK).

[3] A.B. Makhijani and A.J. Lichtenberg, *Environment*, Vol 14, No 5, 1972, p 10.

[4] R.S. Berry and M.F. Fels, *Science and Public Affairs* (Bull.At.Sc.) December 1973, p. 11.

[5] J.C. Bravard, et.al. 'Energy expenditures associated with the production and recycle of metals' ORNL-NSF-EP-24, November 1972 (Available from Oak Ridge National Laboratory, Oak Ridge, Tennessee 37830, USA).

The energy costs of materials

Table 2. Energy costs of metals (kWhth/tonne)

	From ore	From scrap	UK average
Copper	20 000[6]	2 500[7]	12 750
Aluminium	91 000[6]	3 000[7]	27 000
Zinc	20 000[8]	2 500[8]	19 000
Lead	15 000[8]	2 000[8]	7 000
Steel	13 200[a]	6 500[b]	–

Notes:
[a] This is the UK average and is based on 50% scrap. The process used could not operate without a scrap input.
[b] Estimated in reference 5 for a 100% scrap process.

energy costs of these metals produced from ores and scrap metal.

The values for copper and aluminium are in good agreement with those reported by other workers provided that allowance is made for different conventions regarding electricity production (for a detailed comparison see ref. 6). The value for copper produced from primary ores is sensitive to the grade of ore. The values for lead and zinc are in agreement with Makhijani's[3] values provided his results are converted to metric tons and allowance made for the electricity conversion factor.

Plastics and chemicals

A detailed discussion of the energy costs of plastics and other petrochemicals is the subject of Chapter 4. The values given in Table 1 are those appropriate to the UK under the convention that the chemical feedstock is given an energy cost including its calorific value (Chapter 2).

Building materials

There have not been many studies of the energy costs of building materials. MacKillop[9] has given values based on the direct energy consumption of the major industries. His values are correspondingly much lower than those given in Table 1 which include the energy costs of transport, capital equipment and purchased materials. Makhijani[3] has given values for the USA. His values for cement and glass are considerably higher than those in Table 1. In the case of cement a process analysis based on the data in Shreve[10] showed that the US process for producing cement is about 500 kWhth/ton more expensive than the UK process. For glass it is known that the energy cost per ton depends on the type of glass and the age of the furnace being used. The value in Table 1 is the average for the UK production of plate glass and does not apply to other glass products such as milk bottles. The energy cost of bricks given in Table 1 is equivalent to 1·6 kWhth/brick. This is lower than the value deduced by process analysis of 2·0 kWhth/brick.[11] This presumably indicates that the process analysis used data not typical of the average UK plant. The energy cost for wood given in Table 1 corresponds to the energy involved in transporting and cutting the wood to shape. It does not include the calorific value of the wood as a fuel.

Paper

Paper has been included among the materials industries since it is a competitor for a structural material, wood, and some paper products compete with metal and glass products (eg packaging). The value given in Table 1 is exactly the same as that deduced by process analysis from data in Shreve[10] and is within 10% of the

[6] P.F. Chapman, *Metals and Materials*, February 1974, p. 107.

[7] P.F. Chapman, *Metals and Materials*, June 1974 p. 311.

[8] Calculated by methods described in references 6 and 7.

[9] A. MacKillop, *Ecologist*, Vol 2, No 12, December 1972, p. 1.

[10] R.N. Shreve, *Chemical Process Industries*, 3rd Edition, McGraw Hill Co, New York 1967.

[11] Based on publications of the Brick Development Association. See especially: 'The Drying of Bricks' by R.W. Ford. 'Fuels, combustion and heat transfer' by A.D. Aldersley, 1967.

The materials industries

The data deduced from the Census of Production[1] and from the UK Energy Statistics[1,2] can be used to extend the system diagram begun in Chapter 2. As pointed out then the fuel industries themselves consume about 30% of the primary energy input to the UK. However in evaluating the energy costs of materials and commodities this energy loss is incorporated in the energy costs of fuels delivered to final users. This means that the total energy cost of delivered fuels equals the total energy cost input to the fuel industries. This is shown at the top of Figure 1.

The remainder of Figure 1 shows how the delivered fuel is divided between six sectors. The largest fuel consuming sector is the materials sector which consumes 704.8×10^9 kWhth, 29.2% of the energy cost of delivered fuels. The materials sector incorporates the following industries: Mining (Census Report Nos 3–6); Iron and Steel (Census Reports 44–46); Metals (Census Reports 44–49); Chemicals (Census Reports 26–29, 34–36); Building Materials (Census Reports 125, 127, 128, 130, 131); Paper, Rubber and Asbestos (Census Reports 137, 144, 111). The direct fuel input includes the energy cost of the feedstocks delivered to the chemical materials industries. The fuel input also includes fuel used for transport by the industries' own vehicles. The energy input from purchased transport is also shown on the diagram as 10.2×10^9 kWhth. Similarly, the energy cost inputs due to the purchase of capital equipment and other commodities (such as

[12] *United Kingdom Energy Statistics 1973*, HMSO.

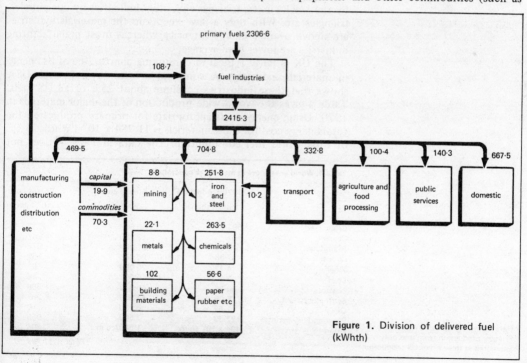

Figure 1. Division of delivered fuel (kWhth)

Table 3. Inputs to the materials industries (units of 10^9 kWhth)

Material industry	Direct fuel and feedstock	Capital equipment	Purchased transport	Purchased commodities	Purchased materials	Gross output
Mining	8·8	1·1	1·5	11·1	0·6	23·1
Iron and steel	251·8	4·4	2·2	11·0	39·5	308·9
Metals	22·1	1·0	0·3	4·0	2·3	29·7
Chemicals	263·5	7·1	2·2	16·4	3·0	292·0
Building materials	102·0	3·5	2·7	20·3	30·5	159·0
Paper, etc.	56·6	2·8	1·3	7·5	21·7	89·9
Totals	704·8	19·9	10·2	70·3	97·5	902·7

explosives) are shown as 19.9×10^9 kWhth and 70.3×10^9 kWhth respectively. The energy cost of the net output of the materials sector equals the sum of all the inputs, 805.2×10^9 kWhth. The division of these inputs and outputs between the six sub-divisions of the sector are shown in Table 3. The gross outputs, column 6 in Table 3, include materials which are purchased by other sub-divisions of the materials sector, shown in column 5 of Table 3.

Earlier it was asserted that the materials industries were amongst the most energy intensive industries in the UK. The evidence for this is shown in Figure 2 which is a plot of *direct energy per £ value added* against *manpower employed per £ value added* for a number of UK industries. The UK average of direct fuel per £ value added is simply total fuel consumption divided by the GDP, equal to 50·6 kWhth/£. This value is shown by the dashed vertical line. Industries to the right of this line have higher than average energy intensity. The materials industries are shown by the black circles, triangles etc., other industries by open circles, triangles etc. With only a few exceptions the materials industries are above average energy intensity whereas most manufacturing industries are lower than average.

The UK is fairly typical in consuming almost 30% of its energy in materials industries. A comparable breakdown for the USA shows that these industries consume about 25% of all US fuels. Table 4 gives the world wide production of the major materials in 1970. Using energy costs appropriate for primary production the total energy cost for these materials is $11\,784 \times 10^9$ kWhth, 21·6% of total world fuel consumption. The data in Table 4 does not

Table 4. World energy use in material production

	1970 Production[a] (million tonnes)	Energy cost (kWhth/tonne)	Energy cost (10^9 kWhth)
Crude steel	592·6	10 500	6 222
Cement	567	2 200	1 247
Aluminium[b]	9·63	91 000	876
Copper[b]	7·58	20 000	152
Zinc[b]	4·77	20 000	95
Lead[b]	3·2	15 000	48
Plastics	30·35	45 000	1 366
Synthetic rubber	5·05	41 000	207
Glass	1·62[c]	6 250	10
Paper and newsprint	127·39	7 500	955
Wood	404×10^6 cu m	1 500/cu m	606
		Total	11 784

[a] Data from UN Statistical Yearbook 1972
[b] Primary production from ores only
[c] Estimated as three times US production

The energy costs of materials

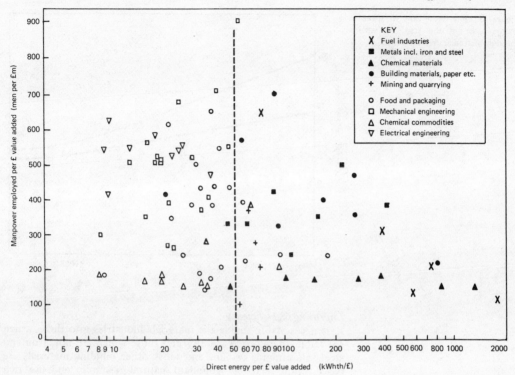

Figure 2. A plot of *direct energy per £ value added* against *manpower employed per £ value added* for various UK industries. The broken vertical line gives the UK average of direct fuel per £ value added.

include fuel consumed in the production of metals from scrap, of building materials (bricks etc) and chemicals since world production figures for these items are not readily available. Were these items included then the proportion of world fuel consumption devoted to materials production might rise to 25 or 30%.

Future demand

Fuel savings are desirable at any time. In a period when fuel prices are rising, when fuel imports cause serious balance of payments problems and the supply of fuel is uncertain, fuel conservation has become a major part of government policy in all industrial nations. The materials industries consume such a large proportion of total fuel that unless attempts are made to reduce the fuel used in this sector then other measures may ,be ineffectual. One obvious solution is for a nation to opt for importing energy intensive materials, thereby passing the problem to someone else. With rising fuel costs this policy is likely to be counter productive in that the trade balance will be made even worse.

There are four other ways in which fuel savings could be made in the materials sector, namely

- improve the fuel efficiency of the industries
- substitute less energy intensive materials for more energy intensive ones
- recycle more materials
- use less materials

Figure 3. The reduction in direct energy cost in the steel industry through technical improvements.

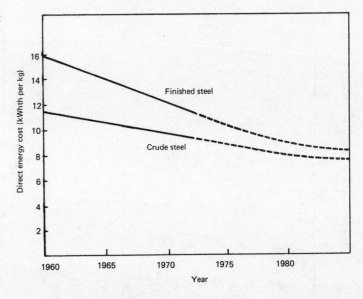

Improving fuel efficiency

It is necessary to divide the materials industries into those which are liable to resource scarcity and those which are not. Iron and steel, aluminium, cement and most other building materials are based upon relatively abundant natural resources and use rich sources of raw materials. For these materials most fuel is consumed in the conversion process, usually a pyrotechnical process. These processes can be improved by better furnace or kiln design, recovery of more waste heat, utilising higher temperatures etc. In all these industries there has been a continuing technical improvement. Perhaps most striking is the improvement in the steel industry shown in Figure 3. (Note this only shows the change in direct fuel consumption). The measures leading to these improvements are many,[13] including an increased use of scrap as a raw material. The extrapolation into the future is based on the estimate of a 10-12% reduction in the energy used per ton of crude steel.[13] The energy cost of finished steel ceases to fall sharply after 1976 since there is a limit to the ratio of crude to finished steel (assumed to be 1·1:1·0 in Figure 3). A similar improvement has been reported[14] for the aluminium industry where the electrical energy required to reduce alumina to aluminium has been reduced from 18 kWh/kg to 15 kWh/kg over a period of 25-30 years. In the next 20 years there may be a further reduction by continued technical improvement or by the development of new processes such as the AARC–Toth[15] process. However, before that time aluminium may have to be produced from anorthosite clays since bauxite reserves are limited. This could lead to an increase in energy cost of about 10%.[6]

In contrast, the materials which are liable to resource shortages have shown a steadily increasing energy cost per ton of product. This arises because as the rich sources are depleted more difficult and lower grade sources have to be worked. In these industries a

[13] Monty Finniston, 'Fewer joules for steelmaking' *New Scientist*, Vol 63, No 905, July 1974, p. 65.
[14] R. Scalliet, 'Consideration of modern cells with pre-baked anodes' in *Extractive Metallurgy of Aluminium: Vol II*, J. Wiley, 1963.
[15] *Metals and Materials* Vol 7, No 9, 1973 p. 370.

large share of the energy cost is usually incurred in the mechanical processing of the raw material prior to reduction or pyro-treatment. In essence the problem changes from being a chemical reaction (performed in furnaces) to a physical entropy change (carried out in complex grinding and flotation circuits). The best example of this type of material is copper[6],* but the remaining metals (lead, zinc, nickel etc.) are subject to similar increases in energy cost. Based on the analysis of the change in energy cost with the ore grade[6] it is estimated that the energy cost of copper has risen by 50% over the past 25 years and will probably be twice today's value in another 20 years.

A different type of resource shortage faces the chemical industry. In this case the energy cost of suitable feedstocks is likely to rise as the easily recovered sources of hydrocarbons are depleted and feedstocks are obtained from coal or tar-sands.* The wood and paper industries also face resource shortages since the growth in demand for these products has exceeded the technical advances in shortening the growing time of trees. Any trend towards using fertilisers or other chemicals to promote tree growth will substantially increase the energy cost of these products.

In conclusion it is apparent that some materials industries offer more room for technical improvement than others. Already resource constraints are increasing energy costs for some materials faster than technical improvements can reduce energy costs. Eventually all materials will be subject to this type of constraint (how long can the UK continue to import high grade iron ore?) but for the next 10-15 years there may be significant savings in the fuel consumed in producing bulk materials. (Note the energy costs may still increase due to increases in the energy costs of fuels*).

Materials substitution
The second method for conserving fuels in the materials industries is to substitute materials with a low energy cost for those with a high energy cost. Unfortunately current trends are precisely the opposite. Aluminium is displacing copper in a number of applications, especially electrical. Plastics are replacing wood, paper and glass.

In comparing alternative materials allowance must be made for the energy cost per unit materials property. For example on a weight basis aluminium is twice as good a conductor as copper so only half a ton of aluminium (45 500 kWhth) can replace one ton of copper (20 000 kWhth). Similarly for milk containers a plastic container may only weigh 0·012 kg (0·54 kWhth) compared to a glass bottle of weight 0·36 kg (2·2 kWhth) and a complete comparison must take into account the number of times each container is used.

Financial criteria may induce substitutions which incur increased fuel consumption. This arises because different materials have different energy coefficients (see Table 1). Thus although it has been reported[16] that concrete oil-production platforms for the North Sea may cost only half as much as a steel platform the energy costs may be the same (compare the total energy per £ value for steel and cement in Table 1). In general there does not seem to be any evidence of the type of substitution required for

* See Chapter 2.

[16] A. McElroy, 'Debate over steel or concrete rigs' *Financial Times* (N. Sea Oil Supplement), June 19 1974, p. 25.

fuel saving, but rather trends and pressures in the opposite direction.

Recycling
The third method for conserving fuels is to increase the fraction of materials produced from scrap rather than primary resources. This is only a realistic option for the metals, paper, rubber and part of the glass industries. Building materials, particularly cement, plaster and wood, cannot be recycled. Plastics could be recycled to a degree but most other chemicals are used destructively or dissipated. Even for those industries which could recycle there is only limited room for improvement. This is because whilst the consumption of a material continues to grow then the quantity of material available for recycling, which was produced some time ago, is only a fraction of present demand. This has been analysed in detail for the copper and aluminium industries[9] and it has been shown that the potential savings were more significant for the balance of payments than for energy consumption. In the case of steel it is estimated that 80% of future demand could come from scrap compared to the present 50%. This would require some modification in the design of products, particularly by large steel consumers such as the motor industry. For the other metals, zinc, nickel, lead, tin etc recycling is already at a high level since there is a substantial price incentive. It is unreasonable to expect that these materials will be recovered from alloys in which they are minor ingredients. Were this attempted then the energy cost of the reclaimed material could be as high as its energy cost from primary sources.

No details of the energy costs of recycling paper, plastics or rubber are known to the author. There is certainly sufficient potential to avoid resource shortages threatening these materials but whether this will lead to a net reduction in energy usage is unknown. In the case of paper, recycling may, as in the case of copper and aluminium, increase UK fuel consumption by displacing imported products.

Reducing consumption
The final method of conserving fuel is the most obvious but also likely to be viewed with the most apprehension. A reduction in the usage of materials raises spectres of unemployment and falling living standards. These are real possibilities, not through any deliberate policy of reduced consumption, but because of continuing economic difficulties in the wake of fuel and commodity price rises. If we had been trying to conserve fuels and materials for many years then a further reduction could be a real threat. As it is there is a lot of slack and wastage in our industrial system and real savings should be possible without serious economic disruption. Some savings will be made as manufacturers try to adjust to higher fuel and materials prices. Others will result from a shift of consumer preferences as relative costs change. For example the increases in running costs of a car will encourage the purchase of smaller, lighter cars requiring less materials for their production and consuming less fuel in use.

There are two factors which, unless carefully considered, could cause serious disruptions. The first stems from the fact that in some situations overall fuel savings could be made by increasing fuel consumption in certain sectors. For example increasing the quantity of materials insulating a house increases both the financial and energy costs of the house, but could lead to net savings. Similarly using aluminium for car engines could increase the financial and energy costs of an automobile, but could lead to a net fuel saving since the automobile would be lighter. In these situations financial criteria discourage the optimum design since the financial costs and benefits acrue to different people at different times. Here there is a need for government to take an over-view and try to encourage optimum fuel use by fiscal and other policies.

The second factor which could lead to unnecessary fuel consumption in the materials industries is concerned with planning. Companies, especially large corporations, have to invest in plant on the basis of their expectations of the future demand for their product. The continued exponential growth in materials consumption has given people confidence in these projections. However, the events of the last twelve months have changed many of the assumptions under which plans were drawn up. The era of cheap abundant fuels has gone forever. The twist comes when agencies and corporations continue to plan on the basis of growth and, then, when demand does not match their predictions, invent uses for their products and advertise so as to *make* their predictions come true. This is in direct conflict with the other pressures induced by the fuel crisis, namely to reduce consumption of fuels and materials wherever possible.

This conflict was apparent in the discussions concerning the number and type of nuclear power stations to be built in the UK. No doubt had the CEGB won the day then its demand curves would have been justified. Similarly if the plastics industry continues to believe in a continued 10% growth per annum[17] then it will happen, even if it means the introduction of disposable toothbrushes and plastic packets of cigarettes. Again this type of problem needs an overview which seeks to formulate an integrated energy and materials policy so as to avoid unnecessary waste.

If plans continue to be made on the basis of extrapolating historical trends then the quantity of fuel consumed in the materials sector will probably rise faster than the rise in GNP. This will come about because the fastest growing industries (plastics and aluminium) are significantly more energy intensive than average. Also materials such as paper and copper which are under resource pressure will tend to be home produced from scrap rather than imported. On this basis, making allowance for the technical improvements discussed above, fuel consumption by the materials industries would increase by 60% by 1985 and by 150% by 2000 even though the consumption of materials over the same period would only increase by 30% and 75% respectively. These increases in fuel consumption could be avoided without hardship, provided plans are made under the new prevailing assumptions, provided fuel and material conservation becomes a major part of good design and provided government action is taken so as to give individuals and industrial organisations the financial incentives to undertake conservation.

[17] 'The Plastics Industry and its Prospects' A report of the Plastics Working Party of the Chemicals Economic Development Committee, National Economic Development Office, (NEDO), HMSO, 1972, p. 89 et seq.

4. An international comparison of polymers and their alternatives

R. Stephen Berry, Thomas V. Long II and Hiro Makino

Following the general analysis of materials' energy costs in Chapter 3, polymers – particularly those used in packaging – are analysed in this chapter from British, American and Dutch data. Four stages are distinguished in the manufacture and processing of polymers: acquisition of feedstocks, manufacture of the monomer, polymerisation, and fabrication of the final articles. Polyethylene, polypropylene, PVC and polystyrene are examined in detail and compared in energy terms with paper and paperboard. Comparisons are also drawn with the energy requirement of metals production.

[1] Chapter 2: 'The energy cost of fuels'.
[2] P.F. Chapman, *Metals and Materials* (February 1974).
[3] J.A. Over (ed), *Energy Conservation: Ways and Means*, (Stichting Toekomstbeeld der Techniek, The Hague, The Netherlands (1974).
[4] R.S. Berry and M.F. Fels, 'The Production and Consumption of Automobiles,' A Report to the Illinois Institute for Environmental Quality, July, 1972.
[5] R.S. Berry and M.F. Fels, *Science and Public Affairs – Bulletin of the Atomic Scientists*, October 1973, p 11.
[6] R.S. Berry and H. Makino, *Technology Review*, Vol 76, No 4, 1974, p 32.

Iron, steel and aluminium

Metals – especially iron and steel and aluminium – compete with polymers in certain applications. A comparison of data on iron and steel making in the UK,[1,2] The Netherlands[3] and the USA,[4,5] and for aluminium production for the same three countries[4,10] reveals that in the UK and the USA, the total energy requirements of hot-rolled steel are about 47 gigajoules/metric tonne (GJ/t) – ie, 47×10^9 joules/metric tonne or 11 900 kWh/short ton – and that the energy requirements for aluminium production are about 250 GJ/t (63 kWh/short ton) in the UK, the USA and The Netherlands. The similarities of the energy requirements among these nations were reassuring, and gave an extra cachet of validity to the calculations. Then the Dutch figure[3] for producing iron and steel appeared: 32 GJ/t, which is 15 GJ/t below the UK-US figures. This of course stimulated a closer look at the apparently similar figures. We concluded that similarities of energy cost figures for the same product made in different countries are liable to be deceptive. Aggregated energy requirements may be very nearly equal and yet be the sums of quite different sets of disaggregated contributions. The illusion of similar totals must not keep us from looking at energy requirements step by step.

Let us illustrate this point by examining aluminium production. The figures are given in Table 1, disaggregated into the three stages of ore extraction, conversion of ore to aluminium oxide, and production of aluminium metal from the oxide by electrolysis in the Hall-Héroult process.[11] The similar values of the sums at the bottom come from two cancelling effects. The energy requirements to convert bauxite ore to alumina vary markedly, with the USA value by far the lowest. In contrast, the figures for electrolysis differ in the opposite direction. The value of 242 GJ/t for the USA represents the national average figure; the values for the UK and The Netherlands are those for the best available technology. If one used the average value of the efficiency of British electrolytic cells, the UK figure for smelting would be 226 GJ/t, and the total for the UK would then be 287 GJ/t. If one used the best available technology for the USA, the new ALCOA process,[12] the US smelting figure would be 170 GJ/t and the total, only 186 GJ/t. Still another difference in the practices of producing aluminium is in the fraction of scrap in the input. About 20% of the US and The Netherlands' productions comes from scrap; in the UK, the fraction is closer to 33%.

Elasticity of demand

It is important to note a relationship between the elasticity of demand for energy and the energy requirements, based on average and most efficient equipment. The output elasticity of demand is the fractional change in output per fractional change in input — in this case, of energy. Thus, if Q is the amount of product, $\Delta Q/Q$ is its fractional change, $\Delta E/E$ is the corresponding fractional change in energy requirement and the elasticity is

$$\left(\frac{\Delta Q}{Q}\right)\left(\frac{E}{\Delta E}\right) = \left(\frac{\Delta Q}{\Delta E}\right)\bigg/\left(\frac{Q}{E}\right)$$

$$\frac{\text{marginal productivity of energy}}{\text{average productivity of energy}} = \frac{\text{average energy requirement}}{\text{marginal energy requirement}}$$

Average and marginal productivities of energy are the inverses of the average and marginal energy requirements.

The marginal energy requirement for producing an additional unit of output is essentially the same as the average energy requirement for the most modern and efficient equipment in use. With this approximation, the elasticity of demand for *increasing* production, which we can call ϵ_{up}, is given by

$$\epsilon_{up} = \frac{\text{average unit energy requirements over all the industry}}{\text{average unit energy requirements with 'best' equipment}}$$

This is not the same as the elasticity of demand for *decreasing* production, which we can call ϵ_{down}, because decreases are generally absorbed in the least efficient (and often oldest) equipment. Rather,

$$\epsilon_{down} = \frac{\text{average unit energy requirement over all the industry}}{\text{average unit energy requirement with 'worst' equipment}}$$

No data are yet available from which we can estimate ϵ_{down}, and we have very little information thus far to find values for ϵ_{up}'s. However, we do, for the first time, have enough to estimate the ϵ_{up} elasticity of demand for energy in aluminium production in the UK and the USA. For the UK, where the new ALCOA process is presumably not yet realistically available, $\epsilon_{up}(E)$ is 287/253 or 1·13.

[7] H. Makino and R.S. Berry, 'Consumer Goods, A Thermodynamic Evaluation of Packaging, Transport and Storage,' A Report to the Illinois Institute for Environmental Quality, June, 1973.
[8] P.F. Chapman, 'The Energy Cost of Producing Copper and Aluminium from Primary Ore,' Report ERG001, Open University, 1973; Chapter 3: 'The energy costs of materials'.
[9] J.C Bravard, H.B. Flora and C. Portal, Energy Expenditures Associated with the Production and Recycling of Metals,' Oak Ridge National Laboratory Report ORNL-NSF-EP-24, 1972.
[10] P.R. Atkins, *Engineering and Mining Journal*, Vol 1974, No 5, 1973, p 69.
[11] A conversion efficiency for electricity of 32% was assumed for UK and The Netherlands and 33% for the USA[2-5]
[12] News release from ALCOA, January 11, 1972.

Table 1. Energy costs of primary aluminium production (Transportation energy is neglected)

Process	UK[8]	The Netherlands[3]	USA[4]
Ore extraction	5 (GJ/t)	5 (GJ/t)	3 (GJ/t)
Alumina production from ore	56	31	13
Aluminium production from alumina	192	196	242
Total	253	232	258

For the USA, we estimate $\epsilon_{up}(E)$ as 258/186 or 1·39, based on the new ALCOA process. Because the historical trend has been toward greater efficiency in the use of energy, we expect to find that, in general, $\epsilon_{up}(E) > \epsilon_{down}(E)$ when we confine ourselves to a narrow technological range of processes. However, wherever we have made substitutions of large energy inputs for large capital or labour inputs, we must watch for situations where ϵ_{up} is actually smaller than ϵ_{down}, due to increases in energy intensity outpacing technological improvements in efficiency.

The aggregate figures for energy used in steel production, likewise, deserve scrutiny. Part of the discrepancy between the low figure for The Netherlands and the UK and US figures given above, occurs because almost all the Dutch steel furnaces are basic oxygen furnaces. If all US production were carried out this way, the energy requirements would drop from 47 GJ/t to 43 GJ/t; another 2 GJ/t of the difference can be assigned to the difference between the coke-oil-gas mixture put into Dutch blast furnaces, as compared to coke alone in the USA. The data available in the open literature appear to be insufficient to account for the remaining discrepancy of 9 GJ/t.

Thus we see that, even for the most easily-analysed, well-documented processes, international comparisons can illustrate quite large differences among practices, and, by implication, indicate useful ways to reduce energy demand.

Polymeric materials

Energy requirements of producing some principal commercial polymeric materials in the USA, the UK, and The Netherlands vary dramatically. We have explored the technological differences underlying these variations to the extent that proprietary interests permit. Energy requirements for fabrication of a few common consumer items are also presented, as well as the costs of paper and paperboard, which are the principal competitors for plastics in the packaging industry.

As Chapman pointed out,[1] there are diverse procedures which one may adopt in energy analysis. Here we are limited by the data for polymer production that is available for the UK[13] and The Netherlands[3,14] to evaluations based on the energy input into the productive system. A more detailed investigation of energy requirements for polymeric materials used by the consumer goods packaging industry in the USA[6,7] calculates both the energy requirements and the free energy requirements.

When we calculate the energy requirements for making an item, we add the inputs of energy from each fuel; the total is a payment required to operate the process. Many fuels contain greater capacities to do work than they furnish by combustion to generate heat. The maximum heat we obtain from a fuel is the parameter known as its heat of combustion. The total potential to do useful work includes the part associated with purity and structure, as well as the part associated with the energy contained in chemical bonds. This total is called the *free energy* ('free' in the sense of useful, *not* in the sense of 'without price'). Thus the free energy is an opportunity cost.[15] It tells us exactly the minimum amount of *energy* that would be required to reconstitute the materials used as fuels in production so that they contained not only their original heat content, but also had the same

[13] Imperial Chemical Industries, Ltd. (private communication).
[14] Private communications.
[15] This point is discussed in more detail by R.S. Berry and T.V. Long, 'The Use of Energy', (submitted for publication).

physical and chemical structure as the original fuel. The free energy cost of a gallon of oil, for instance, is the minimum energy payment that our descendants would have to make to resynthesise the oil from the products of combustion. This could happen if petroleum grew truly scarce and hydrocarbons were desired for their unique, unsubstitutable chemical properties. The actual energy cost of synthesis would always be greater than the free energy; the difference depends on the inefficiencies of the technology used, but the free energy content is the lower bound.

In calculating the amount of primary fuel required for generating, transmitting, and distributing electricity, we have assumed the efficiencies given above for the USA and The Netherlands. A value of 30% has been used for the UK,[13] although a recent study indicates this may be too high.[16] Primary energy costs for producing steam in the USA and The Netherlands are evaluated directly, so that the efficiency of conversion does not enter the calculation. A conversion efficiency of 75-80% was used in the UK study.[13] The US figures for steam correspond to about 82% efficiency, and represent an estimate slightly weighted toward the optimistic side. This corresponds to a fuel input of about 0·474 kWh/lb or 3·75 GJ/tonne of steam. If, as is sometimes done,[6,7] one uses the electrical output potential of the steam as the measure of its energy equivalents, the conversion factor has been estimated to be about 1·1-1·9 kWh/lb or 8·7-15·0 GJ/tonne, corresponding to an efficiency of about 27-35%.

The manufacture of polymers falls into three large steps – acquisition of feedstocks; manufacture of the monomeric material and other inputs to polymerisation; and polymerisation itself. To these, we have to add the energy required to fabricate the final articles. The principal primary resources for polymers are petroleum and, in the USA, natural gas. Petroleum acquisition (US figures) requires about 0·3-0·4 GJ/t. We have assumed, for lack of more data, that this figure is approximately correct for petroleum going to the UK and The Netherlands. It does not reflect any shift to North Sea oil or other energy-costly oil. Natural gas is three times more costly to acquire – over 0·9 GJ/t. Table 2 shows how these totals arise from inputs of electricity, fuel oil, gasoline and natural gas; the data are taken from the US Census data for 1967.[17,18]

[16] P.F. Chapman, 'The Energy Costs of Delivered Energy, UK 1968.' Report ERG-003, Open University, November 1973.
[17] *Census of Mineral Industries 1963 and 1967*, Bureau of the Census, US Dept. of Commerce, Washington, D.C. (1967 and 1971).
[18] W.E. Franklin and R.G. Hunt, 'Environmental Impacts of Polystyrene Foam and Molded Pulp Meat Trays,' MRI Projects 3554-D, Midwest Research Institute, Kansas City, Mo, 25 April 1972.

Note:
The energy supplied by crude petroleum falls in the range 42·7 GJ/tonne of crude (based on international values) to 44·6 GJ/tonne of crude (based on US Census). The energy supplied by natural gas is approximately 54·7 GJ/tonne of gas. Hence the data of ref. 18 implies that the ratio of energy available from combustion (enthalpy of combustion)/ PER is 134-140 for crude oil and 53·1 for natural gas.

Table 2. Fuel inputs to acquire crude petroleum and natural gas[18] (Units are megajoules – 10^6 J – per tonne of petroleum)

Input	Crude petroleum (per tonne)		Natural gas (per tonne)	
	Amount	Energy input (M/J tonne petroleum)	Amount	Energy input (MJ/tonne natural gas)
Primary energy to generate electricity (33% efficiency)	14·0 kWh (from primary source)	55·0	8·4 kWh	33·1
Fuel oil	0·80 gal	132·0	0·62 gal	103·4
Gasoline	0·18 gal	25·3	0·18 gal	26·4
Natural gas	88 cu ft	105·6	363 cu ft	868·0
Total		318		1030

The processing of petroleum to petrochemicals starts with cleaning the petroleum of water-solubles. The water is then removed electrostatically, and the oil is fractionally distilled. The useful fractions are desulphurised, and then sent through catalytic conversion and other chemical transformations to become the useful hydrocarbons for polymer production — ethylene, vinyl chloride, styrene, for example. Outside the USA, the naphtha fraction is used to make ethylene and propylene. The general practice in the USA is to obtain ethylene and propylene by low-temperature distillation of ethane and propane, followed by dehydrogenation.

The manufacture of monomers and polymers is accomplished by highly varied, interwoven and individualised processes. Much of the information is proprietary. Table 3 gives a sample of the total output and the unit energy requirements in the USA for the production stage of most of the important inputs to polymer production. The energy requirements to acquire the feedstocks are not included, so the figures in this table are *not* the total energy requirements. The petrochemicals span a range of about a factor of five, if we omit the alcohols, between about 12 and 60 GJ/t. The greatest total amounts of energy are spent in making olefins, vinyl chloride, acrylate and methacrylate, and styrene, among the monomeric materials. Moreover the *total* energy requirements, which are often given with feedstock fuel values included, can be considerably higher than the figures in Table 3.

Polymerisation is a rather energy-consuming process, despite the fact that many polymerisation reactions actually release heat. A flow diagram for the manufacture of polyethylene is given in Figure 1. Several different processes are used for production of each polymer. They may differ according to catalyst, as in the case of polyethylene and polypropylene formation, or according to the form in which the monomeric material goes into the polymerising process, eg, in

Table 3. Quantities and unit process energy requirements for inputs to polymer manufacture[a]

Material	Quantity (10^3 tonne)	Process energy (GJ/t)
Ethylene	931	16·7
Butadiene	84	16·7
Other olefins	1479	16·7
Acetic anhydride	31·4	28·9
Acrylates and methacrylates	199	56·6
Acrylonitrile	57·2	42·8
Alcohols (except ethanol)	31·7	125·0
Cellulose acetate	54·7	75·5
Urea	93·5	14·0
Vinyl acetate	125	35·6
Vinyl chloride	666	20·1
Formaldehyde	389	10·2
Plasticisers	103	64·8
Phenol	159	14·7
Phthalic anhydride	75·2	19·9
Styrene	707	12·5
Sulphuric acid	309	8·1
Hydrochloric acid	9·4	8·0
Sodium hydroxide	172	5·9
Woodpulp	246	1·5

[a] US figures taken from the US Census of Manufactures for 1967; transportation and feedstock energies are not included

suspension, in an emulsion or in bulk in the formation of poly(vinyl chloride), PVC; or in bulk (without solvent), in solution, in suspension or emulsion for polyethylene. These variations sometimes cause large variations in the requirements for process energy, but not always. The Ziegler (transition metal halide) catalyst method and the metal oxide catalyst method for making polyethylene differ by over a factor of two in their energy requirements, because the metal oxide catalyst method calls for more process steam.[7] Moreover energy requirements can depend on process temperatures – eg, polystyrene formation in bulk is carried out at 230–250°C but only at 20–45°C in solution. In the manufacture of PVC, the emulsion process has high energy requirements because it involves a stage of heat drying.

The production of most polymeric materials is increasing. Outputs for the USA for four recent years are given in Table 4, with the flow diagram for the principal materials in Figure 2. Production figures for a broader spectrum of polymers are given in Table 5 for the USA and UK.

Table 4. USA production (10^6 metric tonnes)

Year	Polyethylene	Polystyrene	PVC	Polypropylene
1962	0·916	0·579	0·712	0·066
1967	1·73	1·08	0·974	0·301
1970	2·65	1·61	1·71	0·469
1972	3·47	2·09	1·95	0·787

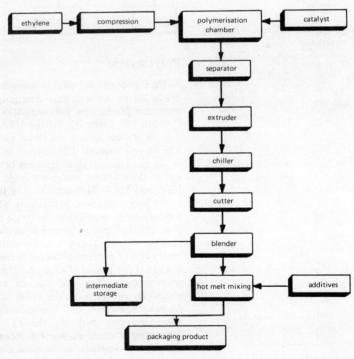

Figure 1. Production of polyethylene for packaging

An international comparison of polymers and their alternatives

Table 5. Production levels of major polymeric materials in the UK and the USA (10^5 tonnes)

	UK	USA (1968)
Plastics, total	12·8	50·2
Regenerated cellulosics	0·11	3·2
Low-density polyethylene resins		13·3
High-density polyethylene resins	all poly-olefins 4·95	7·8
Polypropylene resins		3·3
Polystyrene	1·44	8·7
ABS[a]	n.a.	1·9
Poly(vinyl chloride)	5·06	14·2
Poly(vinyl acetate)	0·37	1·9

Note:
[a] Acrylonitrile-butadiene-styrene

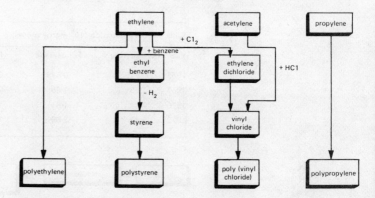

Figure 2. Major plastics in the USA

Polyethylene

Table 6 contains the total process energy to make polyethylene from crude oil or natural gas, disaggregated into steps of acquisition, monomer production, polymerisation and typical product fabrication (film, in this example), for the USA, the UK and The Netherlands. The differences among the three nations' practices are quite striking. The largest relative difference is in the distillation of naphtha from crude oil. The energy differences between using naphtha and natural gas as the ethylene source are large, a factor of 2·5 to 3 between the USA and UK – Netherlands. The largest absolute differences occur in the polymerisation, particularly when UK and Netherlands figures are compared with the metal oxide catalyst method. We have found that most of this difference is associated with the energy used for process steam – partly because almost three times as much process steam is used per unit of output in the USA. (We have used a figure of 0·474 GJ per pound of steam for the USA based on stream flows and process casts. This corresponds to an efficiency of about 0·82, effectively the same as the value of 0·8 used to determine the UK figures.) Some UK and Netherlands chemical plants generate their own electricity and use the by-product steam from the home generating station as process steam. The Dutch operations were designed, it appears, to optimise the use of heat at every possible

a G. Haller, in 'Resource Utilization and Environmental Impact of Alternative Beverage Containers,' presented at the Symposium on Environmental Impact of Nitrile Barrier Containers, July, 1973, from a study carried out for the Monsanto Company. Haller arrives at a figure equivalent to only about 22 GJ/tonne. Of the 40 GJ/t difference, we can account for 10 by differences in prorating among multiple-output processes. With Haller's figure for production of ethylene, the US energy requirement is 120 GJ/t, differing from the European figures entirely because of the energy requirements claimed for the polymerisation step.

b Figures are for low-density polyethylene, in The Netherlands and the UK. For high-density polyethylene, add about 1·8 GJ/t.

c Calculated for the Ziegler catalyst method; if the metal oxide catalyst method is used, the US energy requirements for polymerisation becomes 92·1 GJ/t, and the total, 169 GJ/t, because of the very high steam requirements in solvent recovery and polymerisation.

Table 6. Energy use in the production of polyethylene (GJ/tonne ethylene)

	The Netherlands	UK[13]	USA[6,7]
Production of crude oil or nat. gas	0·3	0·3	0·4
Crude oil → naphtha	1·4	5·7	–
Naphtha → ethylene	25·8	20·5	–
Natural gas → ethylene[a]	–	–	62·2
Ethylene → polyethylene[b]	18·3	18·2	43[c]
Subtotal, for polymer production	45·8	44·7	105·6
Polyethylene → film	3·4[3]	13·7	10·9
Total	49·2	58·4	116·5

stage. Hence some of the difference between the UK and The Netherlands in polymer manufacture may be due to technology, particularly that of heat conduction and insulation.

The total figures are a dramatic illustration of differences in the use of energy to achieve a common end product. The process energy cost of polyethylene film in the USA is almost 2 times that of the UK, and almost 2·4 times that of The Netherlands. If the US manufacturers were to use Dutch practices instead of their own, the manufacture of polyethylene would consume 0·22% of the US energy budget instead of 0·45%, neglecting feedstock energies.

Polypropylene, PVC and polystyrene

Polypropylene follows a pattern qualitatively like that of polyethylene, but with less striking differences. Comparative figures for polypropylene manufacture are given in Table 7.

PVC, whose energy requirements are in Table 8, shows significant international differences, again less than those for polyethylene. We have already seen the differences due to the use of naphtha in The Netherlands and the UK, and of natural gas in the USA. Large absolute and relative differences in PVC manufacture come in the generation of chlorine. A wide range of energy requirements appear in

Table 7. Energy use in the production of polypropylene (GJ/tonne propylene)

	UK	USA
Production of crude oil or nat. gas	0·3	0·4
Crude oil → propylene	33·5	–
Natural gas → propylene	–	56·3
Propylene → polypropylene	28·4	44 approx.
Subtotal	62·2	101 approx.
Polypropylene → film	62·7 (oriented film) or: 17·4 (molded crate)	10·1 (unoriented film)
Total	125·0 (oriented film) or: 79·7 (molded crate)	111 (unoriented film)

a The two major processes using ethylene, based on oxychlorination and on hydrogen chloride oxidation, require 10GJ/t and 20GJ/t, respectively. Two other processes utilise acetylene rather than ethylene; the high-pressure process requires about 10 GJ/t of vinyl chloride, while the low-pressure process generates about 6 GJ/t in the form of a fuel gas.

b By suspension process, with 80% steam efficiency; 19.8 at 32% steam efficiency. If the emulsion process is used, and steam efficiency is 80%, polymerisation requires 14.1 GJ/t (or 16.0, at 33% steam efficiency); the bulk process requires 8.6 GJ/t (or 16.1 at 33% steam efficiency). The emulsion process requires very little energy for steam, but has large fuel requirements for drying, and some energy requirement for refrigeration. The bulk process, by contrast, uses nearly 80% of its energy requirement for polymerisation in the generation of process steam. Ratios of materials assumed are exact chemical equivalents: $C_2H_4 : Cl_2 :$ Vinyl chloride = 0.5 : 0.65 : 1.0.

c Upper figures: total chloralkali process energy assigned to Cl_2, with no regard to simultaneous production of NaOH; Lower figures (parenthesised): 47% of chloralkali process direct energy attributed to chlorine production, 53% to NaOH production.

Table 8. Energy use in the production of PVC (GJ/tonne PVC)

	The Netherlands	UK	USA
Production of crude oil or nat. gas	0.15	0.15	~0.2
Crude oil → naphtha	0.7	5.2	—
Naphtha → ethylene	12.9	9.7	—
Natural gas → ethylene	—	—	31.1
NaCl → Cl_2	33.8 (26.2)	22.6 (17.4)	9.8 (7.6)
Ethylene + Cl_2 → vinyl chloride	13.1	18.6	10-20[a]
Vinyl chloride → PVC	9.1	9.7	9.7[b]
Total	69.8 (62.2)	66.0 (60.8)	60.8-70.8 (58.6-68.6)

the making of vinyl chloride itself, because there are four methods used for this process. Two are based on the addition of hydrochloric acid to acetylene, and two are based on chlorination of ethylene followed by dehydrochlorination of the ethylene so generated.

Polystyrene, as shown in Table 9, exhibits smaller differences among nations than do polyolefins or PVC. The greater energy used in the USA for polymerisation is compensated for, in part, by the larger energy requirements of the British procedure for obtaining benzene from crude petroleum. Overall, polystyrene requires about half the *average* overall energy for plastic resin manufacture in the USA, about 150 GJ/t.

A variety of other aggregated average values of polymers and plastics have been derived, in the neighbourhood of 100 to 150 GJ/tonne. This is clearly the correct order of magnitude for these products. However, as we have seen, the component steps can vary considerably.

Summary

The figures just presented are all *process energy requirements* (PER).

Table 9. Energy use in the production of polystyrene (GJ/tonne polystyrene,[a] excluding transportation)

	The Netherlands	UK[b]	USA[6,7]
Production of crude oil or nat. gas	0.3	0.3	0.4
Crude → benzene	31 approx.	26.3	12.0 (or 26.2[c])
Crude → isopentane	0.3	0.3	0.3
Natural gas → ethylene	—	—	20.9
Crude → naphtha	0.3	3.7	—
Naphtha → ethylene	7.1	7.1	—
Ethylene + benzene → polystyrene	47.9	12.3[b]	44.1 (or 58.3[c])
Total	86.9	50.0	77.7 (or 106.1[c])

a We assume 1 tonne of polystyrene requires 0.85 tonne benzene, 0.96 tonne crude oil, 0.08 tonne isopentane, 0.31 tonne natural gas and 0.30 tonne of ethylene; ie, the chemically equivalent amounts.

b Data presented by H. Smith, 'The Cumulative Energy Requirements of some Final Products of the Chemical Industry,' presented at the World Power Conference, Moscow, August, 1968. The figure of 12.30 GJ/t for preparation of styrene and its polymerisation is an old figure that may be questioned. An independent recent *total* for the UK is 51.6 GJ/t.

c The higher values are based on the energy requirements for production of benzene, as obtained by Haller; see note c, Table 6.

They do not include the energy (strictly, the gross enthalpy of combustion) of the feedstock sequestered in the polymer. The *gross energy requirement* (GER) is the process energy requirement plus the gross enthalpy of combustion of the feedstock. For polyethylene, polypropylene and polystyrene, the feedstocks required for the processes contain about 46 GJ/tonne of product; for PVC the required feedstocks contain about 22 GJ/tonne of product. There is some difference between the GER for olefin feedstocks produced from crude and from natural gas, due to the differences in their heats of combustion and to the PERs for their acquisition. For olefins from crude oil, the GER is greater than the PER by 44·7–47·6 GJ/tonne (46·2 + 1·5), depending on the type of crude and the process losses; for olefins from natural gas, the GER is greater by about 57·4 GJ/tonne. These figures include the best estimates of process losses that were available to us, but would be well worth closer scrutiny.

The information presented here, particularly data released before 1973-74, may be subject to strong bias errors because of inaccessibility of information. The chemical industry, more than any other with which we have dealt, traditionally considered much of its process information to be proprietary. This situation has improved in the past year. Hence older figures are often based on the data from those firms willing to release data, and on census sources which are particularly inadequate in the chemicals field – again, partly because of proprietary reasons, but also partly because of the complexity and diversity of the input-output flow patterns.

The analyses presented here have been validated by comparisons between manufacturers whenever it was possible. Where discrepancies within a single country were large, we have tried to present the differing energy requirements and to explain the origins of the differences, where we could. There are rather wide variations from plant to plant and process to process in the polymer industry, that are considerably greater than our estimate of the uncertainty in the calculations of about 20% in the overall average energy requirements. Because of this variation, we can foresee very large differences between the upward and downward output elasticities, which will only be exposed by a careful and far more detailed analysis of this very complex industry.

Paper and paperboard

The full analysis of paper and paperboard manufacture has already been done for the United States,[6,7] and we shall not undertake to present it here. However, it is useful to compare energy costs of paper and paperboard in the three countries, The Netherlands, the UK and the USA, and see how these materials compete with their most likely alternative, polyethylene, on the basis of their energy requirements.

We illustrate the process of making a typical paperboard product, the folding paperboard box, in Figure 3. The conventions for this figure, as introduced previously,[4,5] follow the recommendations of the International Federation of Institutes for Advanced Study (IFIAS) Workshop *Energy Analysis*, August, 1974. This displays most of the basic quantities and material flows. The energy costs for comparison among nations are given in Table 10. We are inclined to question the figure for The Netherlands taken from reference 3; we suspect that the energy cost is more likely to be in the range of 50

An international comparison of polymers and their alternatives

Figure 3. Energy flow in the production of 1 tonne of folding paperboard boxes

The total is 51·8 GJ/tonne of folding boxes.

Rectangles indicate processes; ovals, amounts in metric tonnes; triangles, amounts of process energy in megajoules (10^6 joules or 10^{-3} gigajoules), and carts, the transportation energy.

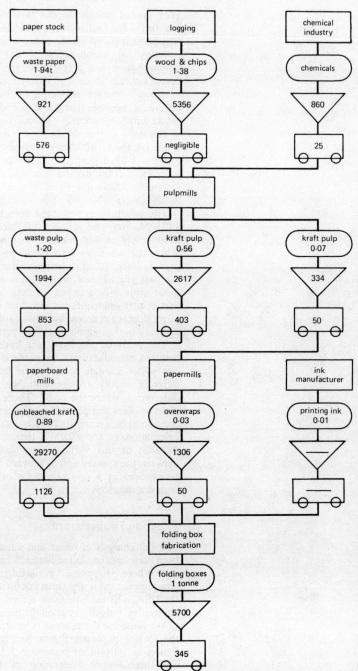

Table 10. Energy consumption in the production of paper and paperboard (GJ/metric tonne)

	The Netherlands	UK	USA
Paper (Kraft)			46·6
Corrugated cardboard	18·8	60·0	40·8
Boxboard			39·0

GJ/t. This point deserves further study.

We can put these figures together into a mundane example, the comparison of the energy input for a paper sack *versus* the energy for a polyethylene sack of the same size and strength. The US figures show the energy cost of a polyethylene sack to be about 1·2 times that of an equivalent paper sack. (This value is lower than the figure of 1·7 given previously[2,6] because the older figure was based on the electrical equivalent efficiency, rather than the direct fuel input, for steam.) The British figures put this ratio at about 0·83, and the Dutch figures, if we accept the low value for the energy cost of paper, give a value of 0·76 because of the extremely low energy requirements for polyethylene production in The Netherlands. Given present practices, polyethylene is a more energy-thrifty material than paper in Europe, but is more costly than paper in the USA. All these figures are based on the same polyethylene-to-paper weight ratio of 2/3, for equivalent containers and on the process energy requirements (PERs) not GERs.

Our analysis suggests that if we are looking for points of significant elasticity of demand in energy within the present context, it is in the US manufacture of polymers and their inputs that we shall find the largest elasticity, certainly not in the Dutch plastics industry, and probably not to the same absolute extent in the paper and paperboard industry. The analysis has also illustrated the importance of making international comparisons, especially on a disaggregated basis, if we are looking for technological optima. International comparisons suggest that energy savings can be analysed and made in specific industries in which the economics are well determined and the risks minimised. In framing both public and private policy, we may turn with considerable confidence to such analyses for guidelines toward readjustment, when confronted with resource constraints.

5. Energy and food production

Gerald Leach

Food production in the industrialised countries is heavily dependent on inputs of energy from fossil fuels. In this chapter, these inputs are analysed with particular attention to the UK. A wide range of data is presented, covering the total energy used and how this relates to factors such as outputs, labour and land. The importance of low energy alternatives for Third World countries is emphasised.

Solar energy alone is not sufficient for food production; other energy inputs are necessary. Recently several studies have asked how much energy and produced answers that are remarkable and rather perturbing. Broadly, they have shown that while most traditional farmers achieve high food yields for each energy unit invested, the industrialised food systems of the West have raised food yields and quality and cut labour usage, but have done so by heavy consumpton of — and dependence on — fossil fuels. Most developed societies now use 7 to 8 units of fossil fuel energy for each food energy unit consumed, or an annual 0·8 tons of oil equivalent per person.

These energy subsidies have helped transform working conditions and living standards in modern societies, especially on the farm. They are also a natural response to a period of high wages and cheap energy. Their emergence is easily explained. Yet they do raise several important policy questions for the future. Not least of these are whether recent trends in the energy-intensive food systems of the West need to be reversed, or can be without harm; whether they are a possible model for the developing world to copy; and if not, what energy-food strategies can do most for the energy- and food-hungry majority of the world's peoples.

The energy analysis approach

This article presents a wide range of data relevant to these questions by focussing on the total energy used in food production and how this relates to other cardinal factors such as outputs, labour and land. Before doing so, three things must be said about the methods and limitations of energy analysis.

First, though it deals with a uniquely important resource, it is merely a descriptive science. It is a complement to, not a replacement for, economic or other valuation systems. This study emphatically rejects the notion often subscribed to energy analysis that because it takes X times less energy to produce an object by method A than by method B, A is somehow X times more worthy.

Second, food production displays such wide variations of outputs and inputs due to climatic, soil, water, management skills and other factors that the energy analysis of food production is at present mostly limited to broad statements about averages. The average data presented here should be treated with this in mind.

Third, and most important, energy analyses themselves vary widely in quality. Tracing all the direct and indirect energy inputs to a crop or national food production system is a formidably demanding task (see footnote on Energy Analysis). Some approximations must be used, and some authors have been a good deal more approximate than others. While data from some authors are not included here on these grounds, those that are given are sometimes not strictly compatible. For this reason the article relies heavily on the author's own study[1] which examines the UK food system in detail with a consistent accounting method and set of data and also applies these to other food production systems. Figures and statements based on this study are not individually referenced here.

The input range

Table 1 shows that energy requirements for a unit of food energy or protein vary by roughly 10 000 times as one spans the entire spectrum of food production systems. For the subsistence and hunter-gatherer communities of the first three entries the Energy Ratio (output/input) is consistently high, thus achieving a traditional aim of agriculture, which is to secure a net energy flow to man. With industrial systems much more energy is needed per unit output, with animal products and sea fishing requiring consistently higher amounts of energy than crops, as one might expect. This last fact does much to explain why for the total farm systems of the UK, USA and Holland the Energy Ratio is always less than one and as low as 0·3. The Table also shows for the UK and Holland a strong trend towards greater energy intensity in agriculture in the last 20 years. When one includes the entire food production and delivery system to the point where food is sold in shops the Energy Ratios in developed societies drop to around 0·2, giving an annual fossil energy requirement of about 24 GJ or 0·56 tonnes oil equivalent *per capita*.

In most developed societies a further 5-10 GJ *per capita* is used in transporting food to the home and in the home for cooking, refrigeration etc. As we shall see later, many rural communities in the Third World consume much greater quantities of fuel in the home than this, bringing their overall energy consumption for food close to the Western level.

On their own these figures have little more than curiosity value. It may be interesting that in the UK over 50 MJ or 1·15 kg oil

[1] G. Leach, *Energy and food production*, IPC Science and Technology Press, Guildford, UK, 1976.

ENERGY ANALYSIS

Energy analysis attempts to measure the 'total energy' required by a process or by the provision of a commodity or service, however indirectly or wherever this energy consumption may have occurred. Total energy refers to any source bearing an opportunity cost and thus includes all fossil fuels at the point of extraction from the ground, wood, and the heat extracted from nuclear fuels, but does not include solar energy. With food systems, muscular work by labour or animals is counted in terms of food energy equivalent (though this is trivial is most industrialised systems) while fuels used for private purposes, such as heating a farmhouse, are excluded. Thus for a tractor ploughing a field the energy requirement (or 'input') includes the gross heat content (enthalpy) of the fuel directly consumed in the tractor; the enthalpies of all fuels required in oil exploration, extraction, shipping, refining, delivery to the farm etc; an appropriate share of the energy used to provide all materials for and to build the machinery and plant in this fuel production-delivery chain; and a similarly comprehensive estimate of the share of the energy required for manufacturing the tractor, for repairs and spares, lubricating oils, etc. In many cases the total indirect inputs exceed the direct, though usually it is only the direct inputs that appear in statistics.

Research methods include use of statistical sources for direct inputs of fuels and electricity purchased by farms and factories; detailed process analyses for major commodities such as fertilisers and tractors; and the use of national input-output tables for estimating the energy equivalents of purchases such as water, machinery, building materials, construction services, transportation, and so forth.

Table 1. Range of energy ratios for food production systems

Farm gate or dockside	Energy out / Energy in	Energy in / Protein out (MJ/kgP)
Chinese peasants 1930s	41	3·6
Tropical crops, pre-industrial*	13-38	4-13
Tropical crops, semi-industrial*	5-10	15-80
Wheat, UK 1970	3·4	42
Maize, USA 1970	2·6	62
Potato, UK 1970	1·6	96
Allotment garden, UK 1974	1·3	58
Rice, USA 1970	1·3	143
Milk, UK 1970	0·37	208
Eggs, UK 1970	0·14	353
Poultry meat, UK 1970	0·10	290
Shrimp fishing, Australia 1974	0·06	366
All fishing fleets, Malta 1970-71	0·04	420
Fisning, Adriatic 1970-71	0·01	1770
Yeast on methanol 1974	—	170
Yeast on N-paraffins 1973	—	195
Winter tomatoes, Denmark (134 MJ/kg)[2]	0·004	14900
Winter lettuces, UK (230 MJ/kg)	0·002	26100
All Agriculture UK 1952	0·46	251
1968	0·34	326
1972	0·35	315
USA 1963[3]	0·87	158
HOLLAND 1950[4]	0·91	
1960	0·53	
1970	0·30	
Total food system to shop door		
UK 1968	0·20	796
USA 1963[3]	0·22	616
1960[5]	0·19	
1970[5]	0·15	
AUSTRALIA 1965-69[6]	0·14-0·20	

* Pre-industrial systems have more than 95% of energy inputs in the form of muscular work by men or animals. With semi-industrial systems the proportion ranges from 10-95% but is usually 40-60%, the remainder being fossil inputs mostly for fertilisers but with some machinery and fuels. All other systems shown are full-industrial, with muscular effort accounting for less than 5% of the total, and usually less than 1%.

[2] B. Elbeck, personal communication, Niels Bohr Institute, Copenhagen, Denmark, 1975.
[3] E. Hirst, 'Food-related energy requirements,' Science, 184, 1974, pp 134-38.
[4] J.M. Lange, De energiehuishouding in de Nederlandse landbouw. Instituut voor mechanisatie, arbeid en gebouwen, Mansholtlaan 10-12, Wageningen, Holland, 1974.
[5] S.S. and C.E. Steinhart, 'Energy use in the U.S. food system', Science, Vol 184, 1974, pp 307-16.
[6] R.M. Gifford and R.J. Millington, 'Energetics of agriculture and food production with special emphasis on the Australian situation', UNESCO – Man and the Biosphere symposium, Adelaide, May 1973.

equivalent is needed to produce a kilogram of eggs at the farm gate, but the more crucial questions arise from comparing these energy data with other key factors of production. This is done first in a broad way, then through a closer examination of the UK.

Energy and labour

An important consequence of the high energy ratios of 'primitive' farming systems is that labour requirements for food supply are not abnormally high, despite popular mythology. With an Energy Ratio of 25 a subsistence farmer need spend only two hours per day on average in order to feed a family of four with a combined food energy intake of 40 MJ per day. This figure is comparable to those of Western societies, where roughly 25-30% of household incomes are spent on food and drink.

Table 2 makes this comparison more explicit by comparing food energy yields per man hour of labour. Most non-industrial cropping systems achieve 10-50 MJ per man hour for raw food delivered to the home. With full-industrial crops this productivity soars to around 3000-4000 MJ per man hour of on-farm labour with food delivered to the farm gate. But these high outputs are then dissipated in two ways. Much of the crop is fed to animals, which reduces the productivity enormously. It is down to 50-170 MJ per man hour on most average UK livestock farms, for example. The second loss occurs when one includes all the other sectors of the food system. The total direct and indirect labour force in UK food production and supply is estimated at close to three million workers, with probably a further one million providing food and feed imports. On this total the labour productivity is as low as 35 MJ per man hour.

This is a notable figure in view of the frequent claims that modern methods allow one farmer to feed 60 or more people. These methods depend on, have allowed and indeed largely caused the vast social changes — including urbanisation and the factory system — which have put large distances between the field and the mouth in every sense and greatly swelled the ranks of non-farm workers in the food system. In fact a food system worker in the UK feeds 'only' 14 to 16 people[7] — a figure that is typical of the middle to upper range for pre-industrial farmers when one counts actual working time. While this comparison does not allow for important differences of climate, food quality, supply reliability, and working conditions, there are also very large differences in the energy requirements as Table 1 shows.

Energy and land

Figure 1 compares energy inputs and outputs per unit of land for a wide range of farming systems. It confirms two important relationships. The first, demonstrated by the sloping diamond of the pre-industrial systems, is that hard work can provide large yields. The highest point of this group, for example, is for traditional Chinese small-holdings of 230 m^2 with labour inputs of 7064 hours per hectare-year and outputs as rice and beans of nearly two tons of protein and 280 GJ per hectare-year. Most of the labour was for collecting dung. Allotment gardens in the UK, with an estimated 14 000 hours of labour per hectare-year yielding 60 GJ and 788 kg of protein as mixed vegetables, also score outputs almost as high as any full-industrial cropping systems on record. Significantly, in both cases high outputs are achieved mainly by virtue of labour intensity and small scale, which allows intensive fertilisation, weeding and double

[7] In the UK average *per capita* food intake is 12·0 MJ per day or 336 MJ per week for a family of four. At 35MJ/man hour, 10 hours of work or one quarter of a man-week is needed to feed the family, not counting any time for food preparation in the home. Alternatively, 56 million are fed by 4 million, giving a ratio of about 14 to 1.

Table 2. Food energy outputs per man hour of farm labour

Agricultural system	Output (MJ/man hour)
Pre-industrial crops	
!Kung Bushmen, hunter-gatherers	4·5
Subsistence rice, tropics	11–19
Subsistence maize, millet, sweet potato, tropics	25–30
Peasant farmers, China	40
Semi-industrial crops	
Rice, tropics	40
Maize, tropics	23–48
Full-industrial crops	
Rice, USA	2800
Cereals, UK	3040
Maize, USA	3800
Full-industrial crops plus animal	
Sheep, cattle, pig and poultry, dairy farms, UK	50–170
Cereal farms, UK (small animal output)	800
UK allotment garden, approx.	4·3
UK food system, approx.	30–35 (all labour)

Energy and food production

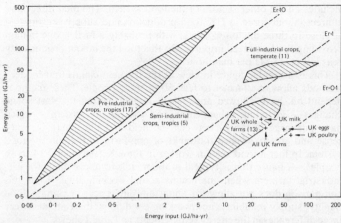

Figure 1. Energy inputs and outputs per unit of land area in food production in the world

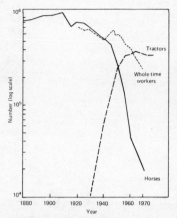

Figure 2. Number of farm horses, tractors, and full-time workers in England & Wales, 1880-1973

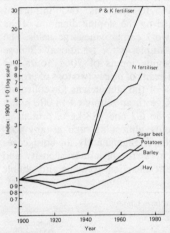

Figure 3. Fertiliser tonnages and crop yields in the UK, 1900-1972

[8] A further 20% of the total farmed area was released by a similar decline in non-farm horses which formed two-thirds of the horse population.

and inter-cropping. However, it is crucial to bear in mind that high yields *and* high labour intensity are rarely a recipe for wealth: for example, at 1974 prices the UK allotment produced a return of only £0·2 (say US$0·5) per man hour.

The second relationship is the more conventional one that large fossil inputs in the form of fertilisers and mechanisation can also give high yields. The cluster of full-industrial crops (cereals, rice, potatoes and sugar beet) have energy yields of 30-80 GJ per hectare-year, roughly three times higher for temperate climates than the majority of pre- and semi-industrial tropical systems.

However, energy inputs are also much higher while — at least on these data — there is a sharp tendency to diminishing returns. As before, the effective yield is greatly reduced by feeding the crops to animals — as shown by the plots for UK livestock farms and animal products.

The UK farm system

A greater insight into food-energy relationships can be gained by looking at the changing patterns of one country. Here I examine the UK, where modern farming methods and a modern industrial-urban food supply structure have evolved to a greater extent than in most countries.

As late as the 1920s UK farming was a pre- or semi-industrial system, with only 10 000 tractors compared to 410 000 today and an average fossil energy input of a mere 100-150 MJ per hectare year compared to 9000 MJ in 1970. Only 6% of farms had a power supply and their combined consumption was less than 1% of present levels.

The transition to full industrialisation occurred very rapidly and mostly in the 30 years since World War Two. Figures 2 and 3 give a rough guide to the scale and pace of these changes. In England and Wales (Figure 2) the number of farm horses declined precipitously, releasing 10% of the total farmed area for food production[8]; whole-time farm workers fell in 50 years from nearly 700 000 to 260 000; and the tractor population rose to about 350 000. At the same time, while crop yields rose to roughly double their 1900 level (Figure 3)

Figure 4. Energy flow for UK agriculture, 1968

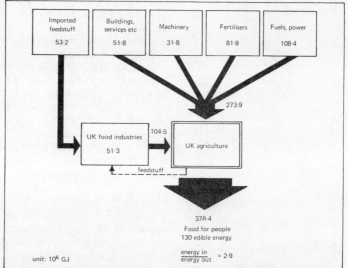

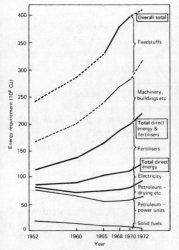

Figure 5. Gross energy requirement for agriculture in the UK, 1952-1972

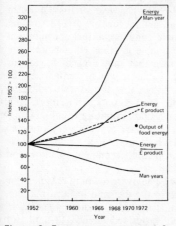

Figure 6. Energy, manpower and £ product, UK agriculture 1952-72

consumption of energy-intensive chemical fertilisers soared. In the same 1900-70 period the total output of animal products roughly doubled, but since the effective area of farmland increased by 28% (largely due to the decline in grazing land for horses) and feed imports also increased, the increase in animal yields per hectare was probably more like 50%.

The effects on energy consumption were substantial. From a very low level at the start of the century, by 1968 the energy input to UK farming had risen to 378 million GJ or 8·76 million tons oil equivalent, equal to 4·6% of UK primary energy consumption in that year. For this investment, among others, farming delivered 130 million GJ of food energy and 1·16 million tonnes of protein for human consumption — enough to feed exactly half the population in energy terms and 62% in terms of protein. A breakdown of these energy inputs is given in Figure 4, while Figure 5 shows rather more approximately how the main classes of energy input changed during the 1952-72 period. The total input increased by 70% from 241 to 410 million GJ or nearly twice as rapidly as gross UK fuel consumption, which rose 40% in the same period.

How did this increase of energy relate to other factors of production? Figure 6 provides some answers. The value added or '£ product' of agriculture rose (in real terms) more or less in step with energy inputs so that energy used per pound sterling remained roughly constant. Since the cost of fuels and power declined in real terms, it became more and more profitable to substitute energy for other basic inputs, especially labour. However, while value added increased substantially, nutritional food outputs increased far less: during the period the output of *edible food for humans* rose by 30% in terms of energy and 35% in terms of protein. Consequently the Energy Ratio declined substantially from 0·46 to 0·35 while the energy to produce a kilogram of protein rose from 251 to 315 MJ or 7·3 kg oil equivalent.

But the most notable changes were in the substitution of energy for manpower. By 1972 each full-time farm worker was backed by a *direct* energy input of 502 GJ or 11·6 tonnes of oil equivalent per year. Counting all part-time workers, directors and the like reduces this to about 180 GJ per man year. Even this lower figure puts agriculture, on this measure, well into the category of heavy industries – in the UK the direct energy per man-year is about 130-140 GJ in engineering and 310 GJ in motor vehicle production. Equally significant, the marginal energy cost of replacing labour appears to have soared. In the early stages of farm mechanisation it often took only 10-20 MJ of energy to save one hour of labour but by 1965-70 this quantity had risen to around 230 MJ. According to the Steinharts[5] the equivalent figure for the USA during the same period was about 720 MJ per man hour saved, though this is not corrected for higher food outputs per hour.

The UK food system

In a developed, urban society such as the UK farming accounts for only a fraction of the total energy required for food supply. Food has to be transported, processed, packed, stored and sold in shops, and in the UK it has to be imported in large quantities. Figure 7 gives an estimate for 1968 of the energy flow for the whole UK food system to the shop door. The total input of nearly 1300 million GJ or 30 million tonnes oil equivalent for a population of 55 million was 15·7% of national energy consumption – though of course a good deal of this energy was 'spent' abroad. The *per capita* consumption of 23·6 GJ is almost identical to an estimate of 23·7 GJ made by Hirst[3] for the USA in 1963, when the food system accounted for 12% of the national energy budget. The Energy Ratios were also much the same at about 0·2 (see Table 1).

These figures do not point to a very energy-efficient food system. Commenting only on the agricultural sector, Blaxter[9] has written that 'taken together, the biological and industrial energetics show both a low biological efficiency of food energy production and a profligate use of fossil fuel energy'. Taking the entire food system only amplifies this comment.

Nor is this a viable system for all people for all time. Copied on a global scale it would demand prodigious quantities of energy – 4000 million people each consuming 23·6 GJ per year of fossil fuels in order to eat (let alone cook) gives an annual fuel bill of 2185 million tonnes oil equivalent or 40% of global commercial fuel consumption in 1972. This figure might be reasonable if other efficiencies were especially high. We have already seen that they are not for labour usage. And nor are they for land. Table 3, which summarises the energetics of the UK food system, shows that the overall biological energy efficiency is only 0·02% while food outputs are, at most, 10·6 GJ per hectare, or only 6·7 GJ per hectare counting rough grazing. Each Briton depends on at least 0·71 hectare for food – ignoring all imports of food and feedstuffs. This is little less than the global average of 1·1 hectare per person counting crop land, permanent meadows and pastures.[10]

Why are these efficiencies so low? The overwhelming reason is the high proportion of animal products in the diet and in farm outputs. The UK farm produces exactly equal amounts of dietary energy in the form of animal products and of crops fed directly to man (see Table 3). Yet while the latter are grown on 1·55 million hectare,

[9] K.L. Blaxter, 'The energetics of British Agriculture', *Biologist*, Vol 22, 1975, pp 14-18.
[10] FAO. *Production yearbook 1973*. UN Food and Agriculture Organisation, Rome, 1974.

Energy and food production

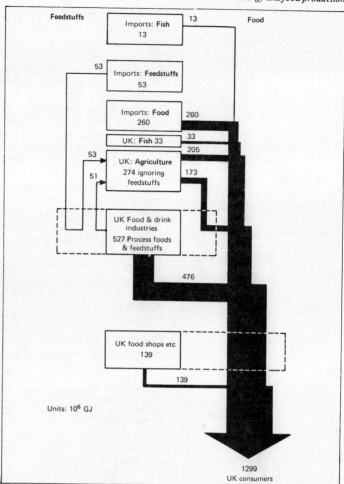

Figure 7. Energy flow for UK food system in 1968

Note: For the food and drink industry as a whole, the total of 527 units breaks down as follows;
direct use of fuels and power	44%
packaging	25%
transport	7%
other (not separated)	24%

[11] K. Mellanby, *Can Britain feed itself?* Merlin Press, London, 1975.

animal production requires 10·25 million hectare of crops and grassland and 1·52 million hectare approximately for imported feedstuffs. Thus the crop sector is seven to eight times more efficient in its use of land to provide food energy than is the animal sector. This is a minimum estimate since it ignores a further 6·65 million UK hectare of rough grazing suitable only for animal raising.

As for fossil fuel efficiencies, a dramatic demonstration of the energy-profligacy of animal production is given by Figure 8. This compares the Energy Ratio for a variety of (average) farms in the UK, all of them producing some crops and some animal products. As the proportion of total energy outputs accounted for by animal products rises from 2% (large cereal farms) to 93% (small specialist dairy farms) the Energy Ratio plummets almost 10-fold. Indeed, with data of this kind it is possible to show that relatively minor reductions in consumption of animal products can give dramatic reductions in the energy inputs and the land requirements for farming, giving for the UK 100% rather than 50-60% self-sufficiency for temperate foodstuffs.[11]

Energy and food production

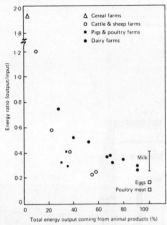

Figure 8. Energy ratios for different types of farms in the UK, 1970-71

Sources: A-C from Blaxter[9]. Remainder from Leach[1].

Energy conservation

Most Western governments are now committed to energy saving policies and energy costs have become a major factor in most production sectors. The food sector is no exception and economies are sure to be made, though the initial response to higher fuel prices has been largely one of passing on higher costs to consumers.

Many minor opportunities exist for energy savings that do little or nothing to alter the structure of farming or other food production sectors. These include greater fuel efficiencies in food processing, tighter control of empty and half-empty freight journeys, reductions in packaging, better insulation and temperature control for glasshouses (which in the UK account for 25% of all on-farm petroleum use), greater care in the use of energy-intensive nitrogen fertilisers with some increase in animal manures, and greater use of 'low tillage' systems in which ploughing and other tractor operations are eliminated or greatly reduced.

The really important routes to fuel economy, however, lie through the *production* of fuels and power on the farm by converting organic wastes and byproducts or deliberately growing fuel crops. For under-developed countries the strategic importance of this marriage of agricultural and energy development is hard to underestimate, and will be discussed after a brief review of possibilities for developed countries.

Developed economies

The theoretical potential for agricultural fuel production or 'biomass conversion' is enormous wherever population densities are relatively

Table 3. Summary of energetics of UK food system, ca. 1968

	Biological flows		10^6 GJ per year
A.	Solar radiation incidence		610 000
B.	Primary production harvested from plants		1 116
C.	Imports of animal feed		104
D.	Edible farm output: crops	65	
	animals	65	
E.	Total edible farm output		130
F.	Food energy consumed by population		261
G.	Primary conversion efficiency (B/A)	0.18%	
H.	System efficiency [E/(B+C)]	10.7%	
I.	Overall efficiency (G x H)	0.02%	
J.	Food energy self-sufficiency (E/F)	50%	
K.	Food protein self-sufficiency	62%	
L.	Food energy output/ha of crops and grass in UK		10.6
M.	L including rough grazing		6.7
	Industrial energy flows		
N.	Energy input to agriculture (and % all UK)	4.6%	378
O.	Energy input to food system (and % all UK)	15.7%	1 300
P.	O counting home-related energy use	(22%)	1 820
Q.	Energy ratio of agriculture (E/N)	0.34	
R.	Energy ratio for whole food system (F/O)	0.20	
S.	R including home-related energy (F/P)	0.14	
			GJ per year
T.	Energy input (O) on *per capita* basis		23.6
U.	Agricultural input/ha for crops and grass in UK		30.7
V.	U including rough grazing		19.5

low, and solar inputs reasonably high, as in the USA. A recent study[12] has shown that a 'practical' development programme for biomass conversion, mainly of farm and forestry wastes, could provide an annual 15 800 million GJ (366 million tonnes oil equivalent) for the USA in the year 2000. This quantity, 24% of 1971 total US energy consumption, equals 60% of the energy output of the largest nuclear programme proposed by the late US Atomic Energy Commission which called for 1400 GW (electrical) installed capacity in 2000 (assuming a 65% load factor). The more modest USAEC target of 850 GW installed in 2000 would produce rather less energy than bio-conversion.

In more densely populated high-energy societies such as the UK, West Germany and Japan the potential for agricultural fuel production is much smaller. In these countries the annual rates of fossil fuel consumption are currently 385, 388 and 278 GJ per hectare of total land surface respectively, which for the UK and Germany are equivalent to just over 1% of solar radiation at the surface. Thus even assuming photosynthetic efficiencies as high as 1%, and complete conversion of crop energies to fuels and power, about 15-20% of the entire land surface would be needed for fuel crops to match present energy use by the food systems alone.

Nevertheless, some agricultural fuel conversions can make a worthwhile contribution. In the UK average yields of cereal straws are 3·5 tonnes per hectare with a gross energy content of 45 GJ. About 3·8 million hectare are devoted to cereals. The conversion of all this material to heat for drying grains or to liquid fuels by hydrolysis or fermentation at 50% efficiencies would yield 85 million GJ of energy, or some 20% of the total energy input to UK agriculture. In practice there may be better uses for straw, such as animal feed, bedding and conversion to paper board. Producing 'biogas' from animal manures is another widely-discussed possibility now being examined in many centres. Taking the UK again, the conversion of all 40 million tonnes of dung from housed cattle would in an efficient system provide around 44 million GJ, leaving nearly all the manure residues for returning to the land as fertiliser. However, high costs (especially for gas storage to match output to demand) appear to rule out wide-scale use of this source at present.

The global challenge

In the underdeveloped world the imperatives of agricultural development are to increase food yields, quality and reliability – and hence the wealth of agricultural communities – without high costs, severe environmental impacts or reduced employment. Carbon copies of Western methods are mostly irrelevant or at worst dangerous. New strategies are needed and in these energy plays a peculiarly important role.

Consider Table 4 which shows how energy is supplied to six 'prototypical' villages in the Third World.[13] These are farming communities. Almost all the human and animal energy is used in food production, including irrigation, and much of the wood, dung and crop wastes are used for cooking. Several striking points emerge:

● When cooking is included energy used in the food system is comparable to that in the West. With these fuels, *per capita*

[12] FEA, *Final task force report on solar energy*, Federal Energy Administration for Government Printing Office, Washington DC, 1975 (GPO No 4110-00012).

[13] A. Makhijani, *Energy and agriculture in the Third World*. Ballinger: Cambridge, Mass., USA, 1975.

Table 4. Energy use in six Third World villages[13]

Village	Gross energy (GJ per capita)				
	Wood, dung, crop wastes	Commercial fuels	Human labour	Draught animals	Total
Mangaon, India	4·2	0·2	3·2	7·9	15·5
Peipan, China	21·1	3·6	3·2	5·3	33·2
Kilombeo, Tanzania	23·2	–	3·2	–	26·4
Batagawara, Nigeria	15·7	0·05	3·0	0·75	19·5
Quebrada, Bolivia	35·4	–	3·5	10·6	49·5
Arango, Mexico	15·1	38·9	3·8	7·6	65·4

consumption for cooking is about 5-7 GJ compared to 1-2 GJ for modern gas stoves and three GJ for electric stoves in the USA[13]

- The cooking fuels are precious resources – dung as fertiliser, crop wastes as manures or animal feeds, and wood as ecological capital. Growing pressures on fuelwood throughout much of the Third World, where annual consumption is often 1-1·5 tonnes per person per year, is creating appalling ecological threats through indiscriminate tree felling with subsequent erosion and creation of deserts. Personal consequences are appalling too. As pressures on wood increase, costs soar and supplies dwindle, many families are forced to scrounge further and further in search of anything that can be burned, even stripping new plantations and the bark off trees.
- Energy supply is overwhelmingly from food or biological sources (the Mexican village excepted) with extremely low efficiencies of use. With draught animals the conversion of fuel to useful work is about 3-5% compared to 25-30% for a tractor. Similarly the conversion of fuels to useful heat in cooking is about 5% compared to 20-25% in a modern gas or electric stove.

In all such communities, adequate power to work the fields and to pump irrigation water (where available) is crucially important for raising yields and avoiding the ravages of drought – and hence for increasing the wellbeing of people. Normally this power is not available because the efficiencies of using energy are so low and the total available energy is limited. Apart from the costly solution of providing more energy from outside by commercial fuels and electrification, the single most urgent need in the food-energy equation is to find cheap ways of harnessing more effectively the energy that is locally available, ie, using it with higher efficiencies. This in turn means providing (storable) energy with high thermodynamic value* – for example, gas or liquid hydrocarbons rather than dung, wood and vegetation.

It is becoming increasingly obvious that bioconversion (and the direct use of solar power) provides the way out of this trap. The skills and technologies are simple, and the fuel sources are widely available, forever renewable (with care) and ecologically inoffensive (with care). Perhaps above all, they are ideally suited to small-scale, self-help, decentralised development which is so relevant to the great majority of the world's poor who still live in scattered rural commities.

Consider what a favoured technology – the conversion of organic matter to biogas (approximately 60% methane) – would mean in general terms for the villages of Table 4. In most of the tropics the year-round solar insolation is in the neighbourhood of 80 000-90 000 GJ per hectare per year (260-290 watts/m^2). Many tropical crops can capture on a year-round basis from 0·5 to 1·0% of this energy without heavy irrigation or fertilisation and about 2% with them. Assuming a low energy capture (0·5%) to allow for energy used in growing and harvesting, and a typical 55-60% conversion efficiency for biogas plant, one arrives at a net yield of about 245 GJ of biogas per hectare year. If this also allows a five-fold rise in the efficiency of using energy for tillage, cooking etc. it would theoretically provide all the gross energy needs for 80 people in Mangaon (India) falling to 25 people in Quebrada (Mexico). At these levels there would be little competition with land for growing food. In practice, more land for crops would probably be made available. Most or all of the organic matter feeding the biogas plant would be crop wastes (and dung) and by eliminating the need for draught animals (let alone much backbreaking human labour) some land would be released for food production. At the same time better tillage methods and irrigation could both increase crop yields and the production of biogas fodder. It is not hard to see how powerful synergistic effects can occur once the stranglehold of the present low energy, low production system is broken.

A similar argument applies to other renewable energy sources which provide concentrated fuels or power. These include liquid fuels obtainable from plant matter by fermentation, destructive distillation or pyrolysis for powering machines such as cultivators or small tractors; electricity from biogas or 'fuel forests'; conventional solar panels to provide hot water and space heating in colder mountain regions; and solar-electric devices such as that proposed by the Meinels of Arizona University for concentrating sunlight onto pipes, storing the energy in molten salts or rocks, and extracting it as required (day or night) to drive turbines to provide electricity at overall conversion efficiencies as high as 25%.[14]

The development and diffusion of energy devices of these kinds throughout the rural areas of the Third World is an enormous challenge. Perhaps above all the challenge is to the broadness and subtlety of our vision. Many of these devices have been costed by the conventional calculus of economics and have been found either to be wanting or only marginally attractive compared to more conventional supply technologies. The question is whether the conventional calculus with its high rate of discounting the future and its failure to catch many of the most relevant factors in its net, is the most appropriate guide. No one has yet thoroughly explored the multitude of consequences and transformations that developments of this kind could bring about — not least on food production, on rural incomes, on personal wellbeing and self-respect, on the invigoration of village life, and on the mass migrations to the exploding cities of the Third World; in short, on the whole development process. Adequate energy may not be the only essential in the mosaic of development out of poverty, but at the very least such a broad-based exploration seems long overdue.

[14] W.E. Morrow, 'Solar energy: its time is near', *Technology Review*, December 1973, pp 31-42.

6. The energy cost of goods and services:

an input-output analysis for the USA, 1963

David J. Wright

The input-output table is a matrix showing the flows of products – in money terms – between the various industries of a national economy. Using the 1963 US government input-output tables which divide the economy into 363 industries, energy costs of all commodities produced in the USA are derived. The results are presented as energy costs per unit of money value, which are appropriate for comparing money and energy costs for different commodities, and also energy costs per physical unit, which are useful in obtaining a relation between physical production and demands on primary energy sources.
In the next chapter, a separate independent study using 1963 and 1967 input-output data for the USA shows how different assumptions lead to different results.

The views expressed in this article are those of the author and do not necessarily coincide with those of the UK Department of the Environment. The article is Crown Copyright ©1974.

The recent 'energy crisis' was not due to the fact that energy is a depletable resource, nor due to the fact that there are applications in which the substitutability of energy is limited. It was due to the fact that the producers were able to control the supply and hence the price of energy. However, any of these three reasons would justify a close appraisal of how much energy is used in the production of the various goods and services in our economic systems. This can be done by constructing an energy budget for each production process and results have been obtained by Bravard and Portal,[1] Chapman,[2,3] Leach and Slesser,[4] Smith[5] and Slesser.[6] Not only the direct inputs of energy to the process are counted, but also the secondary requirement of energy in the production of the other raw materials input. Another way of calculating the energy requirements of different commodities is from published government statistics. A variety of such sources is used by Makhijani and Lichtenberg[7] and in the present article we use the input-output table. This table is a matrix showing the flows of products between the various industries of an economy.

Obtaining energy costs from input-output tables

The economy is divided into a number of industries and these are listed as both the rows and columns of the input-output matrix. The entries in the column corresponding to any selected industry give the direct requirements from each of the other industries to produce one unit of that industry's output. If the input-output matrix is denoted by A, then the element A_{ij} gives the requirement from industry i to produce one unit of the output of industry j. It is important that the matrix is square, as then, knowing these direct requirements, one can use the same matrix to find out what was needed to produce them. These one can call the secondary requirements, and it can be seen that they correspond to the matrix A^2. Similarly, one can calculate the tertiary requirements corresponding to A^3 and so on. Finally, the total requirements matrix, B, which represents all these indirect and direct inputs is given by

$$B = A + A^2 + A^3 + \ldots = (I-A)^{-1} - I$$

Thus a matrix inversion is all that is needed to obtain the total requirements matrix from the direct requirements one, and this inverted matrix is usually published along with the matrix A.

As regards energy requirements, everything has, in effect, been traced back to primary energy inputs in the matrix B, taking into account inefficiencies of the fuel conversion industries. Thus, if we identify those rows of B that correspond to inputs of primary energy (coal, crude oil, natural gas, nuclear and hydro-electricity), we can obtain the total primary energy requirement of any commodity by reading down the corresponding column of the matrix B.

The input-output tables are, of course, published in money terms and so we obtain the requirement:-

x dollars of primary energy per 1 dollar of cement   (1)

for instance. Knowing the prices of the different forms of primary energy, this can be converted to:

y kWh of primary energy per 1 dollar of cement   (2)

and, knowing the price of cement, we obtain:

z kWh of primary energy per 1 ton of cement   (3)

However, many commodities are insufficiently homogeneous to be measured by a physical unit such as 1 ton, and we can obtain only (2) above, eg

y kWh of primary energy per 1 dollar of farm machinery.

Each of these measures is useful in different ways. Some applications are as follows:

(1) gives an idea of the short-term effect of changes in the price of energy on the costs of commodities. Comparing (2) for different commodities it can be seen whether there is any disparity between money costs and energy costs. (3) gives the relation between physical production and demands for primary energy. Results of type (1) can simply be read off the input-output table, and in the present article we concentrate on those of types (2) and (3).

Calculation of energy costs of US products

Input-output tables cannot be derived from the economic statistics normally available to governments in countries with market economies. The derivation of all the inter-industry flows requires a special census which is not conducted annually. Moreover, obtaining the input-output table from the census data is itself a time-consuming process with the effect that present input-output tables are, at best, five years out of date by the time they are published. The most recent UK table is for 1968[8] and this divides the economy into 90 industries, whereas for the US a bigger table is available, 363 industries square, for 1963[9]. The present author[10] has previously calculated energy costs from the UK tables; this chapter concentrates on the USA. In fact, results for the USA are easier to obtain as foreign trade (which introduces a complication into the calculation) is a much smaller proportion of GDP in the US than in the UK.

[1] J.C. Bravard and C. Portal, *Energy requirements in the production of metals* (Oak Ridge Laboratory Report, USA, 1971)

[2] P.F. Chapman, 'The energy costs of producing copper and aluminium from primary sources' (Open University Report ERG 001, UK, 1973)

[3] P.F. Chapman, *The energy costs of producing copper and aluminium from secondary sources* (Open University Report ERG 002, UK, 1973)

[4] G. Leach and M. Slesser, *Energy equivalents of network inputs to the food production process* (Strathclyde University Report, UK, 1973)

[5] H. Smith, 'Cumulative energy requirements of some products of the chemical industry' *Transactions 20, Section E, World Energy Conference*, 1969

[6] M. Slesser 'Energy subsidy as a criterion for food policy planning' *Journal of the Science of Food and Agriculture*, Vol 24, Nov 1973

[7] A.B. Makhijani and A.J. Lichtenberg, 'Energy and well-being', *Environment*, Vol. 14, No. 5, June 1972, p. 10

[8] *Input-Output Tables for the UK, 1968*, Central Statistical Office of UK, (Her Majesty's Stationery Office, 1973)

[9] *Input-Output Structure of US Economy: 1963*, US Department of Commerce, 3 vols (US Government Printing Office, 1969)

[10] D.J. Wright, 'The natural resource requirements of commodities' Paper presented at conference on energy costing held at Imperial College, London, July 1973

Table 1 gives results of type 2 above in units of kWh/$ for each of the 363 industries in the US 1963 input-output table. The energy unit kWh has been chosen, not because of any electrical connotation, but as a well-defined physical measure avoiding the tons of coal equivalent in which primary energy is usually measured. The money unit is the 1963 $.

Some details relevant to this calculation are as follows:

(a) *Prices of primary energy*

Prices of the different forms of primary energy are obtained from the 'Statistical abstract of the US'[11] and checked against production figures and the financial data in the input-output table itself. Imports are also treated as an input of primary energy to reflect the importation of energy and also the energy used abroad to manufacture exports to the USA. An average figure is taken as the energy cost of imports. These prices are taken to apply to purchases of primary energy by any industry so that concessionary prices allowed to certain industries are a source of error in the calculation.

(b) *Prices of final products*

The unit of value assigned to any commodity is the factory gate price measured in 1963$. Thus transportation costs (energy and monetary) incurred by the producer directly, or indirectly, to transport his supplies and his supplier's supplies, etc, are included, but the costs of distribution to the final buyer appear separately under 'wholesale and retail trade' (items 69.1 and 69.2) and 'transportation' (industries 65.1-65.7).

(c) *Capital depreciation*

The inter-industry flows in the input-output table include repairs to, but not replacements of, capital. So if one regards depreciation as replacements of capital, then this energy cost is ignored in the analysis. We see below that there is a straightforward relationship between money and energy value of capital so that, given the monetary value of capital depreciation, one could easily translate this to an energy cost and some authors have done this for certain products. However, depreciation rates are not readily available for each of the 363 industries separately and so this energy flow is excluded from the present analysis.

(d) *Labour, profits and solar energy*

These are also energy inputs to the production process which have been excluded from the results in Table 1. They could be included but there is some arbitrariness as to the appropriate stage to do this. For instance, is it the physical energy contributed by the labourer to the production process, or the energy content of the food he eats, or the primary energy needed to produce all the goods and services he enjoys? Similarly, with solar energy in agriculture, should we include the total energy incident on the plants, or just that used in photosynthesis, or an opportunity energy cost of not covering the farm with solar cells? Different uses of the results give different answers to these questions. Excluding these three inputs of energy means that the results are

[11] 'Statistical Abstract of the United States', (US Department of Commerce, Washington, DC) (Published Annually)

Table 1. Energy intensities of US products in kWh/1963$

Ref. no		(kWh/1963$)
(1.01)	Dairy	18.5
(1·02)	Poultry	22·9
(1·03)	Meat	18·6
(2·01)	Cotton	21·4
(2·02)	Grains	25·4
(2·03)	Tobacco	20·9
(2·04)	Fruits	16·1
(2·05)	Veg. sugar	19·0
(2·06)	Oil crops	24·0
(2·07)	Forest prods.	16·2
(3·00)	For. & fish prods.	38·6
(4·00)	Ag. for. & fish serv.	16·2
(5·00)	Iron ore mining	47·4
(6·01)	Copper ore mining	21·5
(6·02)	Non-ferr. ore mining	38·2
(7·00)	Coal mining	257·0
(8·00)	Crude oil & nat. gas	33·5
(9·00)	Stone & clay mining	28·3
(10·00)	Chem. & fert. mineral mining	31·2
(11·01)	Constr. resid. buildings	14·9
(11·02)	Constr. non resid. buildings	16·6
(11·03)	Constr. public utilities	20·9
(11·04)	Constr. highways	30·8
(11·05)	Constr. other	27·8
(12·01)	Maint. resid. build.	13·0
(12·02)	Maint. other	14·4
(13·01)	Guided missiles	7·4
(13·02)	Ammunition	16·5
(13·03)	Tanks	20·6
(13·04)	Sighting & fire control	10·4
(13·05)	Small arms	21·9
(13·06)	Small arms ammo.	14·4
(13·07)	Other ordnance	13·6
(14·01)	Meat Products	18·0
(14·02)	Creamery butter	21·4
(14·03)	Cheese	21·6
(14·04)	Con. & evap. milk	21·7
(14·05)	Ice cream	17·0
(14·06)	Fluid milk	17·9
(14·07)	Canned sea foods	21·8
(14·08)	Canned specialities	17·5
(14·09)	Canned fruits & veg.	20·0
(14·10)	Dehydrated food	17·1
(14·11)	Pickles & sauces	18·9
(14·12)	Packaged fish	23·3
(14·13)	Frozen fruits & veg.	18·0
(14·14)	Cereal preparations	20·0
(14·15)	Animal feeds	19·9
(14·16)	Rice milling	21·2
(14·17)	Wet corn milling	30·1
(14·18)	Bakery products	15·8
(14·19)	Sugar	46·3
(14·20)	Confectionery	24·6
(14·21)	Alcohol	9·8
(14·22)	Soft drinks	16·3
(14·23)	Flavouring	18·8
(14·24)	Cottonseed oil mills	23·3
(14·25)	Soybean oil mills	25·4
(14·26)	Vegetable oil mills	50·3
(14·27)	Fats & oils	27·3
(14·28)	Roasted coffee	52·3
(14·29)	Cooking oils	23·5
(14·30)	Manufactured ice	22·6
(14·31)	Macaroni & spaghetti	16·5
(14·32)	Food preparations	22·1
(15·01)	Cigarettes	8·2
(15·02)	Tobacco stemming	20·1
(16·01)	Broadwoven fabric	24·4
(16·02)	Narrow fabric	19·9
(16·03)	Yarn	24·4
(16·04)	Thread	22·7
(17·01)	Floor covering	26·5
(17·02)	Felt goods	19·2
(17·03)	Lace goods	24·7
(17·04)	Paddings	20·2
(17·05)	Textile waste	22·9
(17·06)	Coated cord	21·9
(17·07)	Tyre cord	27·0
(17·08)	Scouring	37·5
(17·09)	Cordage	48·5
(17·10)	Textile goods	59·4
(18·01)	Hosiery	13·9
(18·02)	Knit apparel	14·2
(18·03)	Knit Fabric	21·4
(18·04)	Apparel other	13·0
(19·01)	Curtains	16·0
(19·02)	House furnishings	21·9
(19·03)	Textile products	18·2
(20·01)	Logging camps	23·6
(20·02)	Sawmills general	23·8
(20·03)	Hardwood	23·9
(20·04)	Special sawmills	33·5
(20·05)	Millwork	15·3
(20·06)	Veneer & plywood	23·6
(20·07)	Wood structures	16·7
(20·08)	Wood preserving	28·4
(20·09)	Wood products	23·4
(21·00)	Wooden containers	20·7
(22·01)	Wood furniture	14·2
(22·02)	Upholstered furniture	14·6
(22·03)	Metal furniture	21·8
(22·04)	Mattresses	18·4
(23·01)	Wood office furn.	15·3
(23·02)	Metal office furn.	19·3
(23·03)	Public build. furn.	18·7
(23·04)	Wood partitions	13·6
(23·05)	Metal partitions	21·0
(23·06)	Venetian blinds	20·3
(23·07)	Furniture & fixtures	17·3
(24·01)	Pulp	51·3
(24·02)	Paper	47·2
(24·03)	Paperboard	41·8
(24·04)	Envelopes	20·5
(24·05)	Sanitary paper	26·8
(24·06)	Wallpaper & board	37·5
(24·07)	Paper products	26·8
(25·00)	Paperboard boxes	24·7
(26·01)	Newspapers	14·2
(26·02)	Periodicals	14·1
(26·03)	Book printing	11·9
(26·04)	Misc. publ.	11·8
(26·05)	Commercial printing	18·4
(26·06)	Business forms	15·9
(26·07)	Greetings cards	12·6
(26·08)	Misc. printing	9·3
(27·01)	Industrial chemicals	61·9
(27·02)	Fertilisers	36·3
(27·03)	Agric. chemicals	47·0
(27·04)	Misc. chem. products	42·0
(28·01)	Plastics materials	42·2
(28·02)	Synthetic rubber	60·3
(28·03)	Cellulosic fibres	45·0
(28·04)	Organic fibres	27·9
(29·01)	Drugs	14·0
(29·02)	Cleaning preparations	27·1
(29·03)	Toilet preparations	16·1
(30·00)	Paints	40·3
(31·01)	Petroleum refining	375·0
(31·02)	Paving mixtures	109·0
(31·03)	Asphalt	83·6
(32·01)	Tyres	25·9
(32·02)	Rubber footwear	15·0
(32·03)	Reclaimed rubber	24·1
(32·04)	Misc. plastic products	21·9
(33·00)	Leather tanning	19·0
(34·01)	Footwear	16·1
(34·02)	Footwear except rubber	11·1
(34·03)	Other leather products	12·7
(35·01)	Glass	18·8
(35·02)	Glass containers	23·1
(36·01)	Cement	11·2
(36·02)	Brick	39·7
(36·03)	Ceramic tile	28·9
(36·04)	Clay refractories	29·7
(36·05)	Clay products	37·6
(36·06)	Plumbing fixtures	18·5
(36·07)	Food utensils	18·1
(36·08)	Porcelain elec. supplies	15·1
(36·09)	Pottery products	16·6
(36·10)	Concrete block	30·1
(36·11)	Concrete products	24·4
(36·12)	Mixed concrete	39·9
(36·13)	Lime	100·0
(36·14)	Gypsum products	24·3
(36·15)	Cut stone	22·4
(36·16)	Abrasive products	18·2
(36·17)	Asbestos products	27·5
(36·18)	Gaskets	18·1
(36·19)	Minerals	28·9
(36·20)	Mineral wool	20·2
(36·21)	Non-clay refractories	32·8
(36·22)	Non-metal mineral prod.	23·5
(37·01)	Blast furnace	68·9
(37·02)	Iron & steel foundries	21·1
(37·03)	Iron & steel forgings	41·2
(37·04)	Primary metal products	21·2
(38·01)	Primary copper	32·9
(38·02)	Primary lead	29·1
(38·03)	Primary zinc	39·7
(38·04)	Primary aluminium	49·2
(38·05)	Primary non-ferr.	58·1
(38·06)	Secondary non-ferr.	12·3
(38·07)	Copper rolling & drawing	25·6
(38·08)	Alum. rolling & drawing	32·9
(38·09)	Non-ferr. rolling & drawing	24·8
(38·10)	Non-ferr. wire drawing	21·5
(38·11)	Alum. castings	24·0
(38·12)	Brass castings	19·0
(38·13)	Non-ferr. castings	20·1
(38·14)	Non-ferr. forgings	27·6
(39·01)	Metal cans	35·4
(39·02)	Metal barrels	35·6
(40·01)	Metal sanitary ware	20·7
(40·02)	Plumbing fittings	20·1
(40·03)	Heating equip. not elect.	18·4
(40·04)	Structural steel	31·4
(40·05)	Metal doors	20·8
(40·06)	Plate work	28·7
(40·07)	Sheet metal	28·3
(40·08)	Architectural metal	21·5
(40·09)	Misc. metal	33·4
(41·01)	Bolts, etc.	23·9
(41·02)	Metal stampings	25·6
(42·01)	Cutlery	20·0
(42·02)	Hand tools	19·9
(42·03)	Hardware	18·3
(42·04)	Engraving, etc.	18·3
(42·05)	Misc. wire prod.	41·2
(42·06)	Safes	20·9
(42·07)	Steel springs	35·7
(42·08)	Pipe fittings	20·3
(42·09)	Collapsible tubes	20·1
(42·10)	Metal foil	24·8
(42·11)	Metal products	23·0
(43·01)	Steam engines	21·4
(43·02)	Int. comb. engines	14·7
(44·00)	Farm machinery	22·9
(45·01)	Constr. machinery	18·3
(45·02)	Mining machinery	17·3
(45·03)	Oil field machinery	17·7
(46·01)	Elevators	17·3
(46·02)	Conveyors	17·4
(46·03)	Hoists	20·5
(46·04)	Industr. trucks	17·3
(47·01)	Mach. tools cutting	14·8
(47·02)	Mach. tools forming	19·7
(47·03)	Special dies & tools	14·7
(47·04)	Metalworking machinery	15·2
(48·01)	Food prod. mach.	17·8
(48·02)	Textile machinery	20·2
(48·03)	Woodworking mach.	18·0
(48·04)	Paper ind. mach.	21·9
(48·05)	Printing mach.	15·0
(48·06)	Special mach.	19·4
(49·01)	Pumps	15·5
(49·02)	Ball bearings	23·1
(49·03)	Blowers	15·7
(49·04)	Indust. patterns	9·3
(49·05)	Power transmiss. equip.	17·5
(49·06)	Furnaces	16·8
(49·07)	General ind. mach.	16·1
(50·00)	Machine shop prod.	13·0
(51·01)	Computing machines	9·0
(51·02)	Typewriters	11·9
(51·03)	Scales	15·3
(51·04)	Office mach.	16·8
(52·01)	Aut. vend. mach.	16·6
(52·02)	Laundry equip.	18·7
(52·03)	Refrig. mach.	16·6
(52·04)	Measuring pumps	17·4
(52·05)	Serv. ind. mach.	15·9
(53·01)	Elec. meas. instr.	10·8
(53·02)	Transformers	23·2
(53·03)	Switchgear	12·7
(53·04)	Motors & generators	16·1
(53·05)	Industrial controls	12·0
(53·06)	Welding apparatus	20·6
(53·07)	Carbon products	40·7
(53·08)	Elect. ind. apparatus	16·6
(54·01)	House cooking equip.	18·6
(54·02)	House frig.	18·8
(54·03)	House laundry equip.	18·8
(54·04)	Elec. housewares	17·8
(54·05)	Vacuum cleaners	13·6
(54·06)	Sewing machines	13·1
(54·07)	Household appliances	18·8
(55·01)	Electric lamps	9·9
(55·02)	Lighting fixtures	25·4
(55·03)	Wiring devices	20·3
(56·01)	Radio & TV sets	16·5
(56·02)	Phonograph records	15·6
(56·03)	Telephone apparatus	10·8
(56·04)	Radio & TV commun. equip.	8·7
(57·01)	Electron tubes	13·1
(57·02)	Semiconductors	13·7
(57·03)	Electronic components	13·4
(58·01)	Storage batteries	18·3
(58·02)	Primary batteries	13·8
(58·03)	X-ray apparatus	15·8
(58·04)	Engine elec. equip.	15·5
(58·05)	Elec. equip.	15·9
(59·01)	Bus bodies	20·3
(59·02)	Truck trailers	19·8
(59·03)	Motor vehicles	17·4
(60·01)	Aircraft	8·8
(60·02)	Aircraft engines	15·0
(60·03)	Aircraft propellers	13·9
(60·04)	Aircraft equip.	10·6
(61·01)	Shipbuilding	15·5
(61·02)	Boatbuilding	15·0
(61·03)	Locomotives	14·3
(61·04)	Railroad cars	27·6
(61·05)	Motorcycles	35·6
(61·06)	Trailer coaches	16·9
(61·07)	Transportation equip.	26·4
(62·01)	Eng. & sci. instruments	11·1
(62·02)	Mech. meas. devices	12·3
(62·03)	Aut. temp. controls	12·5
(62·04)	Surgical instruments	16·8
(62·05)	Surgical appliances	13·3
(62·06)	Dental equipment	17·9
(62·07)	Watches	11·4
(63·01)	Optical instr.	18·7
(63·02)	Opthalmic goods	18·8
(63·03)	Photo equip.	17·0
(64·01)	Jewellery	31·6
(64·02)	Mus. instr.	17·8
(64·03)	Games	17·5
(64·04)	Sporting goods	14·4
(64·05)	Pens	17·9
(64·06)	Artifical flowers	43·5
(64·07)	Buttons	18·1
(64·08)	Brooms	19·3
(64·09)	Floor covering	20·2
(64·10)	Morticians goods	15·7
(64·11)	Signs	18·0
(64·12)	Misc. manu.	20·5
(65·01)	Railroads	17·3
(65·02)	Local highway pass. trans.	27·2
(65·03)	Motor freight trans.	18·5
(65·04)	Water trans.	45·4
(65·05)	Air trans.	43·5
(65·06)	Pipe line trans.	29·4
(65·07)	Trans. services	5·2
(66·00)	Communications	5·5
(67·00)	Radio & TV broadcast	6·1
(68·01)	Electric utilities	104·0
(68·02)	Gas utilities	197·3
(68·03)	Water & sanitary serv.	25·2
(69·01)	Wholesale trade	11·5
(69·02)	Retail trade	8·2
(70·01)	Banking	4·3
(70·02)	Credit agencies	16·2
(70·03)	Security brokers	6·4
(70·04)	Insurance carriers	8·9
(70·05)	Insurance agents	7·1
(71·01)	Owner-occupied dwellings	2·7
(71·02)	Real estate	10·5
(72·01)	Hotels	9·7
(72·02)	Repair services	12·5
(72·03)	Barber & beauty shops	4·0
(73·01)	Misc. bus. serv.	8·3
(73·02)	Advertising	11·6
(73·03)	Misc. prof. serv.	7·0
(75·00)	Auto repair	11·0
(76·01)	Motion pictures	13·8
(76·02)	Amusements	6·8
(77·01)	Doctors & dentists	3·7
(77·02)	Hospitals	9·2
(77·03)	Other medical serv.	12·1
(77·04)	Education	13·2
(77·05)	Non-profit organisations	10·0
(78·01)	Post office	7·2
(78·02)	Federal elec. util.	239·0
(78·03)	Commodity credit corp.	—
(78·04)	Other fed. gov.	19·8
(79·01)	Loc. gov. pass. transit	14·9
(79·02)	State elec. util.	108·0
(79·03)	Other state ent.	22·2
(80·01)	Direct imports	—
(80·02)	Transferred imports	—
(81·00)	Business travel	24·8
(82·00)	Office supplies	19·0

appropriate to an investigation of demands made on terrestrial energy sources, solar energy being used to the extent to which it is today, and people enjoying consumption for its own sake rather than to facilitate their contribution to the production process.

(e) *Feedstocks*

It is not possible to distinguish in the input-output table between, for instance, petroleum used as a feedstock and petroleum used to provide process energy. Hence both are automatically included, giving a very high energy costing for plastics, etc. The figures therefore include the opportunity cost of not using the feedstock energy elsewhere.

(f) *Technology assumption*

All figures are inevitably based on the technology of US industry in 1963. This is important to bear in mind for industrial outputs produced by a mixture of technologies. For instance, the energy cost of aluminium reflects a weighted average of the energy cost of recycling and that of extraction from bauxite. This sort of factor is probably more important than the changes in technology since 1963, since energy consumption and real GDP have risen at approximately the same rate since then (at just over 4% in the USA from 1963 to 1971). Only with the recent rise in primary energy prices can one expect technical change towards energy saving.

To summarise, the factors included and excluded are listed in Table 2.

Interpretation of results

We now describe three different areas of application of these results:

(1) *Energy cost of capital investment*

Table 1 shows the 'energy intensities' (kWh/$) of the different commodities produced in the USA. The 'flow' of energy in final products at these different 'intensities' can be represented by the

Table 2. Factors included and excluded in energy costings

Included	Excluded
Mining, quarrying of raw materials	
Metal extraction, refining, fabrication	
Transportation of intermediate products between industries	Transportation and distribution of finished product to final buyer
Lighting, heating of factories and offices	
Et cetera at all previous stages of production process	
Repairs to capital	Replacements of capital
Feedstocks	
	Labour and Profits
	Solar Energy
Energy lost in fuel conversion	

histogram in Figure 1. The vertical axis shows the money cost of purchases by final buyers (personal consumption, capital investment, government purchases, exports). The average energy intensity for the whole economy is 23·4 kWh/$ but there is quite a spread about this value (standard deviation 52·9 kWh/$) because of some commodities, which we shall consider below, whose energy intensity is too high to fit on the horizontal axis of Figure 1. Within this histogram, a separate histogram is drawn for capital investment, and shows that, although 115 of the industries make contributions to this, the spread of energy intensities of capital investment is relatively small. (Mean value = 16·5 kWh/$, standard deviation = 4·4 kWh/$). This is similar to results obtained from an analysis of UK input-output tables by the present author[10] where a narrow range of energy intensities of capital goods was also observed (mean value 44·1 kWh/1963£ = 15·8 kWh/1963$, standard deviation 3·3 kWh/1963£ = 1·2 kWh/1963$). One of the advantages of the input-output method of deriving energy costs, over the more detailed process analysis approach, is in rapidly identifying broad classes of products, such as capital goods, which have roughly the same energy intensity. One can then say with some degree of certainty that a $1 million investment programme will require roughly $16·5 \times 10^6$ kWh of primary energy, without going into the details of what buildings and machinery, etc, are involved.

(2) Energy flows in foreign trade

Another straightforward result from Table 1 and Figure 1 is to identify those commodities whose energy intensity is much higher than average. Excluding fuels, whose energy intensity is obviously high, the most striking are some foods, textiles, paper, chemicals, plastics, paints, asphalt, cement, metals, transportation. These energy intensity figures are particularly relevant to trade, as they show whether we are gaining or losing energy when we exchange commodities of equal money value. Of those listed above the most important in trade are paper, chemicals, iron and transport. Chemicals take oil as a feedstock and it is a preferred

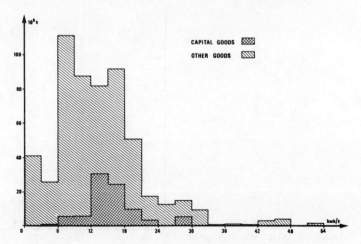

Figure 1. Energy flow at different 'intensities' in US 1963

fuel for transport, but paper and iron can be produced without it. Hence if we hypothesise that oil prices will rise relative to the cost of other primary fuels to 12-15 times the 1963 figure, it will be possible to save the same amount of energy by reducing exports of paper and iron as by increasing imports of crude oil (for the same foreign exchange cost). In 1974 oil prices were 4-5 times their 1963 values.

A more comprehensive analysis of energy flow in foreign trade is shown in Figure 2. As mentioned above, foreign trade is more important for the UK than for the USA, and so the histograms are similar to Figure 1 but derived from the UK results.[10] In Figure 2 one histogram shows energy flowing from UK in all exports excluding fuels (mean = 41·8 kWh/1968£, standard deviation = 29·8 kWh/1968£, total energy flow = 336 x 10^9 kWh) and the other histogram shows the comparison for imports (mean = 51·7 kWh/1968£, standard deviation = 30·7 kWh/1968£, total energy

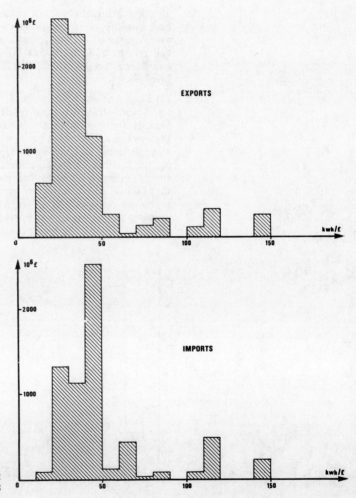

Figure 2. Energy flow at different 'intensities' in UK foreign trade 1968

flow = 339 x 10^9 kWh). Thus the energy flows balance, but imports are, on average, more energy intensive than exports (the difference is significant at the 2·5% level).

(3) *Energy cost of services*

It is often said that, as per capita income rises, the consumption pattern tends more towards services than goods which, being less energy intensive, allow some slackening in the rate of growth of demand for energy. Certainly this tendency has not had an appreciable effect on the energy/money ratio for different nations in recent years. Cross-section and time series analyses[12] of energy comsumption per capita/GNP per capita show this ratio tending to an upper asymptote as income per head rises rather than declining. Table 1 throws some light on this point to the extent of identifying the energy intensity of services. Financial and personal service industries, health, education, etc, all have an energy intensity of about 10 kWh/$, ie just under half the figure for the whole economy. So, although the energy intensity of services is certainly less than average, it is not so small that the current trend towards a 'service economy' produces a noticeable effect on overall energy consumption.

Energy costs in physical terms

So far we have been concerned with energy costs per money value of the product, and now we turn to energy costs per physical unit, ie the measure of type (3) described earlier. This is only available for those commodities which are sufficiently homogeneous to be measurable in physical units and a list is given in Table 3. As stated above these are figures for the average technology employed, so that for metals the energy cost given is lower than the cost of extraction from ore, because some is produced by (less energy intensive) recycling. Another reason for some of the figures being too low is that concessionary prices for bulk purchases of fuels have not been taken into account.

The figures in Table 3 can be expected to be less accurate than those in Table 1 because no industry produces a completely homogeneous output. Also the input-output approach is more useful for obtaining overall results for a broad range of commodities as described above, and where interest centres on a specific product a detailed process costing may be called for. For instance, the very high figure for paint in Table 3 corresponds to a similar high figure obtained from the UK input-output tables,[10] and probably indicates that a process energy costing is advisable to identify where energy savings can be achieved.

Summary

Energy costs of all commodities produced in the USA in 1963 have been derived from an input-output table. All intermediate inputs to the production are traced back to the primary energy needed to produce and transport them. Other authors have performed energy costings of individual processes and obtained more detailed results, whereas the input-output approach is more

[12] L.G. Brookes, 'More on the output elasticity of energy consumption', *Journal of Industrial Economics*, Vol 21, No 1, Autumn 1972, p. 83

Table 3. Energy costs in US in 1963 units kWh/kg unless otherwise stated

No.	Commodity	Energy cost
(1·2)	Poultry	7·32 (per kg live weight)
(2·1)	Cotton	15·7
(2·3)	Tobacco	26·4
(5·0)	Iron ore	0·504
(14·2)	Butter	27·4
(14·3)	Cheese	20·3
(14·6)	Milk	7·35
(14·14)	Flour	2·36
(14·16)	Milled rice	4·09
(14·19)	Sugar	5·74
(14·24)	Cottonseed oil	5·33
(14·25)	Soybean oil	4·98
(14·28)	Coffee	41·9
(15·1)	Cigarettes	0·0370 kWh/cigarette
(20·6)	Plywood	1·41 kWh/ft^2
(24·1)	Pulp	7·83
(24·2)	Paper	17·02
(27·2)	Fertilisers	8·63 (per kg plant nutrient)
(28·2)	Synthetic rubber	35·8
(30·0)	Paints	199·7 kWh/gal.
(32·3)	Reclaimed rubber	6·10
(37·1)	Pig iron	4·34
(37·2)	Steel (ex-foundry)	3·27
(37·3)	Steel forgings	6·39
(38·1)	Copper	22·1
(38·2)	Lead	7·14
(38·3)	Zinc	10·5
(38·4)	Aluminium	24·4
(38·8)	Aluminium sheet	30·4

appropriate for producing results for a range of commodities, eg capital goods, imports, exports, etc. Input-output tables are usually available in published form whereas data for process costing are sometimes less easy to obtain. However, input-output tables are at least 5 years out of date because of the time currently taken to carry out the required census and then compile the table. This applies to the technology used, not to the prices. The availability of input-output tables facilitates comparisons between countries although this is not attempted in this chapter. Also the input-output approach can readily be applied to any other natural resource, or combination of resources, besides primary energy.

We have presented above energy costs per unit of money value (Table 1) which are appropriate to comparing money and energy costs for different commodities, and also energy costs per physical unit (Table 3) which are useful in obtaining a relation between physical production and demands on primary energy sources. It is difficult to estimate the accuracy of these figures but anything better than 10-15% would be optimistic.

Acknowledgement

The author would like to express his thanks to former colleagues at the Systems Analysis Research Unit of the Department of the Environment, London SW1 (where this work was done) for many helpful discussions.

7. The energy cost of goods and services:

an input-output analysis for the USA, 1963 and 1967

Clark W. Bullard, III and Robert A. Herendeen

Using 1967 US input-output data which divide the economy into 357 sectors, this study determines the primary energy inputs to all sectors, taking account of the fact that energy is sold at different prices to different customers. Possible uncertainties in the model and updating methods are discussed, and comparisons are drawn with the study described in Chapter 6.

The work was supported by the National Science Foundation.

When you consume anything, you are consuming energy. The emerging art of 'energy analysis' seeks to determine *how much* energy is required to provide goods and services. Here we describe a method based in part on static input-output economic analysis which we have applied to the United States economy (357 sectors) for 1967. The method allows explicit treatment of the flow of energy involved in the flow of goods across regional boundaries, which has key significance for the question of energy self sufficiency.

There are many reasons for wanting to quantify the energy cost of goods and services. Initially we were motivated by an interest in energy conservation and the potential for saving energy through substitution of products and services. Chapman[1] has pointed out that observation of energy costs of natural resources may provide a firm basis for estimating recoverable reserves, taking full account of the Second Law of thermodynamics, and accounting for the absolute scarcity of free energy reserves. Taking the concept one step farther, Wright[2] has calculated the natural resource requirements for a number of consumer goods.

In a simultaneous, but independent study, Wright[3] estimated energy costs of goods and services, using a technique similar to ours, based on input-output data for 1963. His results, however, differ substantially from ours due to two simplifying assumptions. The methodological differences will be noted below, and the effect of these assumptions on numerical results are discussed later.

General discussion of method

The basis is the idea of 'conservation of embodied energy'. This says that the energy burned or dissipated by a sector of the economy (say a steel mill) is passed on, embodied in the product.[4] Applying this to every sector yields the following picture: primary energy is extracted from the earth, is processed by the economy, and ultimately gravitates to final demand (ie, personal and government consumption and exports). The method also yields the energy intensity, that is, the embodied energy per unit of output, for each economic sector.

Before proceeding, we must add a caveat. This approach covers the whole spectrum of consumer goods and services. In return for the gift of a large body of data from economics, we as energy analysts have had to make several simplifying assumptions, as well as using data that are eight years old. We thus obtain results which are exhaustive, and which are applicable to large scale questions, but which are less useful for very detailed, micro questions (for example, the energy cost

[1] P. Chapman, 'No overdrafts in the energy economy,' *New Scientist*, May 17, 1973, pp 408-410.
[2] D. Wright, 'The natural resource requirements of commodities,' *Applied Economics*, v 7, 1975, pp 31-39.
[3] Chapter 6: 'The energy cost of goods and services: an input-output analysis for the USA, 1963'.
[4] This is in addition to energy from the earth (for a primary energy sector) or in the form of feedstocks.

Figure 1: Conservation of embodied energy for an economic sector.

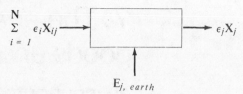

of different building materials). For the latter specific process analyses should be more accurate.[5,6,7,8]

In Figure 1, we apply 'energy balance' to an economic sector.

X_{ij} is the transaction from sector i to sector j

X_j is the total output of sector j

ϵ_j is the embodied energy intensity per unit of X_j

$E_{j,\,earth}$ is that energy extracted from the earth by sector j, and is non-zero only for primary energy sectors. (All quantities are measured for a standard time period.)

We assume that the energy embodied in inputs to sector j, plus the energy burned in that sector, is passed on as part of j's output. This 'energy balance' yields one equation for each of the N sectors; we then solve for the ϵ_j.

So far we have not specified the units of the transactions X_{ij}; the validity of the energy intensities depends on the choice. Ideally, we would like to use physical units (tons of steel, cubic yards of concrete, etc), since these would presumably serve as good linear allocators. Adequate physical data do not exist at the 357-level of detail, so instead we rely mainly on dollar transactions data from Reference 9. This is the data base of an input-output analysis (I-O) of the US economy carried out every 5 years by the US Dept of Commerce.

Dollar data are inferior to physical, being more subject to economies of scale. Reliance on monetary data for energy transactions effectively assumes energy is sold at the same price to all users. Since this assumption is most questionable for the USA, we use physical data (Joules)[10] exclusively. The transactions table X_{ij} is thus in mixed units. This method differs from that of Reference 3 which assumed that energy was sold at the same price to different users.

Attention must be paid to imports. Ideally, we would like to remove them from the energy flow diagram in order to calculate energy intensities for domestic technology, and then reintroduce them to account for their embodied energy flow into the country. This is complicated by the existence of two kinds of imports in the economic data we use.[9] First, transferred – also known as competitive – imports which have domestic counterparts, such as steel. Second, directly allocated – non-competitive – imports, which do not, such as bananas. Transferred imports of steel are added to the output of the domestic steel sector. Since the inputs needed to produce that steel are not counted, we remove transferred imports from the output as shown

[5] R. Berry and M. Fels, 'The production and consumption of automobiles, Report to the Illinois Institute for Environmental Quality, July 1972.

[6] B. Hannon, 'System energy and recycling: A study of the beverage industry,' CAC Doc No 23, Center for Advanced Computation, University of Illinois, March, 1973.

[7] B. Commoner, M. Gertler, R. Klepper, and W. Lockeretz, 'The effect of recent energy price increases on field crop production costs,' Report, Center for the Biology of Natural Systems, Washington University, St Louis, Missouri, December, 1974.

[8] A. Makhijani and A. Lichtenberg, 'Energy and wellbeing,' *Environment*, vol 14, No 5, June, 1972, pp 10-18.

[9] *Input-Output Structure of the US Economy: 1967*, vols. I-III, US Department of Commerce, 1974. Published by the US Government Printing Office. The data are available on tape from Bureau of Economic Analysis, US Department of Commerce. Additional explanation is found in the February, 1974, *Survey of Current Business*, and in *Definitions and Conventions of the 1967 Input-Output Study*, unpublished but available from the Bureau of Economic Analysis.

[10] *1967 Census of Manufacturers, 1967 Census of Mineral Industries*, US Department of Commerce, were basic documents. A complete explanation is D. Simpson and D. Smith, *1967 Direct Energy Transactions*, Technical Memorandum No 39, Center for Advanced Computation, University of Illinois.

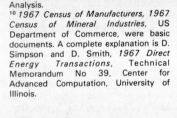

Figure 2: Energy balance for a domestic sector. P_j is transferred imports.

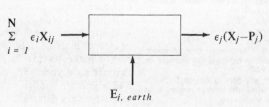

in Figure 2. Here again our method is different from that of Reference 3 which did not reduce gross outputs by the amount of imports.

No similar correction is possible for those directly allocated imports which are inputs to domestic sectors. (We simply do not know the energy intensity of jade, teak, or bananas.) Fortunately, these imports are relatively small; in 1967 directly allocated imports sold to producing sectors were worth $3·8 billion* against transferred imports of $22·6 billion.[11] We therefore neglect them in calculating energy intensities. For a nation with larger imports this would introduce significant error.

Once the energy intensities are obtained, we can treat transferred imports as if they embody the same energy as their domestic counterparts. (One way to justify this is to note that they would require this much energy if they were manufactured here.) Directly allocated imports must be assigned an approximate energy intensity.

The US economy, or that of any nation, may be viewed as receiving energy in three ways:

1. Primary energy (coal, crude, gas, hydro, nuclear) from the American (or nation's own) earth.
2. Imported energy (for the USA, almost exclusively petroleum), with an associated embodied energy penalty due to losses in extraction, refining, etc, carried out abroad.
3. The energy embodied in imported non-energy goods.

Computational details

For the present we assume that only one kind of energy is extracted from the earth. The approach can be extended to several kinds of energy as well.

We assume that each sector is in energy balance, from Figure 2:

$$\sum_{i=1}^{N} \epsilon_i X_{ij} + E_{j\ earth} = \epsilon_j (X_j - P_j) \qquad (1)$$

In matrix notation we have

$$\epsilon = E_{earth} (\hat{X} - \hat{P} - X)^{-1} \qquad (2)$$

where

ϵ is the row vector of energy intensity coefficients
$\hat{X}$ is a diagonal matrix with gross outputs, X_j on the diagonal
$\hat{P}$ is a diagonal matrix with transferred imports, P_j on the diagonal
X is the transactions matrix
E_{earth} is a row vector with one non-zero term (corresponding to the one assumed primary energy sector).

It is helpful to normalise with respect to domestic output

$$\epsilon = E_{earth} (\hat{X} - \hat{P})^{-1} (I - X(\hat{X} - \hat{P})^{-1})^{-1} \qquad (3)$$

and to define a matrix A of domestic technological coefficients:

$$A = X(\hat{X} - \hat{P})^{-1} \qquad (4)$$

Then Equation (3) can be rewritten

$$\epsilon = e(I - A)^{-1} \qquad (5)$$

---

* 1 billion = 1000 million

[11] By definition, all transferred imports are inputs to producing sectors. In contrast, most directly allocated imports are sold directly to final demand ($14·43 billion in 1967). The latter have no bearing on the calculation of energy intensities.

where e is a vector whose elements are zero except for the energy sector; that element is unity.

To arive at Equation (5), we have assumed only that there exists a vector ϵ which results from applying Equation (2) to the base period data. Our intent is to make ϵ more useful by requiring that it apply, in the stated linear fashion, Equation (1), to *any* set of transactions that might occur. For example, if twice as much of commodity j is produced, twice as much embodied energy is implied. This linearity assumption, which equates average and marginal energy intensity, is a weakness of the method. There are two approaches to this problem:

The first is to apply a sufficient condition: let **A** be constant, independent of scale and time. This is the assumption of standard input-output (I-O) analysis.

The second is to apply only a necessary condition, that ϵ is constant. This leads to a large set of equations relating the X_{ij}.

The first approach requires the specification of more information than the second; for the purpose of obtaining only ϵ it is too strong. Hence at this point it is *not* necessary to apply the usual I-O assumption.

For the case of the transactions table X_{ij} expressed in both energy and dollar flows, the units of ϵ are Joules/Joule for energy sectors, and Joules/dollar for non-energy sectors. This is illustrated for an example three-sector economy in Appendix A (p.80).

We have identified the energy intensity ϵ_j as the energy embodied in – needed to produce, directly and indirectly – a unit of product j. This interpretation implies some double counting if we sum the energy embodied in the output of all sectors (for the same reason that adding sector dollar outputs exceeds the gross national product). It is easily shown, however, that ϵ_j is also the energy needed to produce a unit of product j delivered to final demand. Summing the energies necessary to produce all final demands yields just the energy inputs to the economy. This justifies the concept of final demand as the final sink for all energy. (See Appendix B, p.81).

The ϵs obtained above may be used to compute the energy impact of an arbitrary final demand.

Extension to several kinds of energy

In the 357-sector breakdown of the US economy, there are five energy sectors (coal, crude oil and gas wells, refined petroleum, electricity, and gas utilities). One might wish to obtain either energy intensities for a certain energy type, or the total primary energy required (ie, the sum of the coal, crude oil and gas, and hydro and nuclear power).

For the first purpose one can treat the energy sector in question as if it receives its domestic output from the American earth, even if it is a secondary energy producer like electricity. One therefore solves Equation (3) with a non-zero entry for E_{earth} only in the relevant energy sector. Doing this for all five energy sectors effectively converts E_{earth} into a matrix (5 x 357, with 5 non-zero entries), and now ϵ becomes ϵ (5 x 357)

$$\epsilon = E_{earth} (\hat{X} - \hat{P} - X)^{-1}$$

This is also illustrated in Appendix A.

The total primary energy coefficient is a linear combination of the respective single primary energy coefficients. Thus

$$\epsilon_{total\ primary, j} = \epsilon_{coal, j} + \epsilon_{crude + gas, j} + \alpha\, \epsilon_{electricity, j}$$

α is a factor to account for the electricity produced from hydro and nuclear sources. Refined petroleum and fossil electricity are secondary energy types and are omitted to avoid double-counting. The actual value of α depends on the convention one uses for energy 'costing' of these sources. In this chapter we use $\alpha = 0.6165$, based on a heat rate of 11 133 Btu/kWh and the fact that in 1967, 18.9% of US electricity was from hydro or nuclear sources. This follows the prevailing US convention of costing them according to the fossil fuel technology they replace. Very likely a different convention is appropriate for a nation with a high percentage of hydroelectricity.

Energy intensities for the US economy, 1967

From Reference 9, and from independent determination of the energies used by the sectors in the base year,[10] we have obtained enough data to apply Equation (3) to the US economy in 1967. We stress that for most sectors we have explicitly accounted for the fact that energy is sold at different prices to different customers (eg cheap electricity to aluminium smelters.) Results are available for the five kinds of energy and for total primary energy.* For the non-energy sectors, the intensities are in thousands of Joules per dollar, while for energy sectors the units are Joule per Joule. Subject to the conditions mentioned below, they can be applied to a variety of problems. We give a few example applications below, but most have been published elsewhere.

Some of the potential limitations of the results derive from data problems, but others derive from economic conventions used in computing the I-O data base:

- I-O data are subject to inaccuracies from lack of complete coverage of an industry, restriction of information for proprietary reasons, and use of different time periods for data on different sectors. Also, errors in **A** may generate disproportionate errors in $(I-A)^{-1}$.
- The use of dollars rather than physical units to express physical dependencies is less than perfect. For example, aggregation can combine in the same sector two processes whose energy intensities differ widely. And, as we mentioned, economies of scale may be implicit in the dollar data, whereas there would be little or no corresponding effect in physical terms.
- There is a problem with secondary products. The definition of an I-O sector is based on the establishment rather than activity. For example, if those establishments which produce primary aluminium also produce aluminium castings (amounting to less than 50% of total sales), the primary aluminium sector is credited with the summed output. The secondary output is transferred to the aluminium castings sector, ie treated as a sale. The corresponding inputs are not transferred. This means that the dollar output corresponding to production of these aluminium castings has been counted twice, but the energy only once. The fraction transferred varies from sector to sector, so that a

* A table of results referring to the 1967 US economy is available from the authors.

correction is required. In Reference 12 an approximate correction was used. There is an exact method,[13] but inadequate data to implement it. The results here incorporate no correction for secondary products.

- A problem arises in capital goods; these are not considered part of the inter-industry transactions but are listed as sales to final demand. Conceptually, we would consider the energy to make a steel forming press owned by an auto manufacturer to be as valid an energy contribution as that used to make the steel in the auto itself, but this is not compatible with the primary data source[9] which defines the system boundary consistent with the definition of GNP. Calculations to incorporate capital flows are also described in Reference 13, but they are not applied to the results here.
- Final demand is measured in producer's, not purchaser's prices. Since two of the I-O sectors are wholesale and retail trade, it is possible to make the conversion, including the energy requirements implied in the markup (as has been done for the automobile, below). For direct purchases by consumers, it is desirable to convert beforehand to purchaser's prices.
- Input-output coefficients change with time, yet we hope to use the results to predict the consequences of hypothetical future consumption patterns. Can one quantify their loss of reliability with time? This is a major point, for which much work is needed. Our feeling is that our results are most sensitive to changes in direct energy use coefficients, which may change faster than others due to fuel substitutability and the potential for energy conservation.
- As mentioned, the assumption of linearity is equivalent to equating marginal and average energy intensity, which is questionable.
- The assumption that foreign technology is as energy intensive as domestic may be wrong and will introduce error into analysis of imports.

Comparison with other results

The results obtained by Wright[3] are for 1963, so are not directly comparable with those presented here, which are for 1967. We have, however, also done a similar calculation for 1963[14], comparable to Wright's. As a basis for comparison, we used the difference in the two values (for total primary energy) divided by our result. Treating the intensities from each sector, as independent and of equal weight, we find Wright's figures average 12% lower than ours, with a mean deviation of 23%.[15] Thirty-four of the intensities differ by more than 50%.

That energy intensities calculated by our domestic base method average higher is due to the fact that Wright's approximation admits imported goods at zero energy cost; ours costs them as if they were produced domestically. Errors are greatest in those sectors where imports are a large fraction of sector output.

The deviation not explained by imports is due to Wright's admitted assumption that energy is sold at a uniform price to all consumers. While this assumption may be valid for some countries, it is certainly not true for the USA where declining block rate structures are

[12] R. Herendeen, 'An energy Input-Output matrix for the United States, 1963: User's Guide,' CAC Doc No 69, Center for Advanced Computation, University of Illinois, March, 1973.
[13] C. Bullard and R. Herendeen, 'Energy impact of consumption decisions', *Proceedings of the IEEE, Special Issue on Social Systems Engineering*, March, 1975.
[14] C. Bullard III and R. Herendeen, 'Energy Cost of Consumer Goods 1963/67,' CAC Doc No 140, Center for Advanced Computation, University of Illinois, November, 1974.
[15] Not standard deviation, which would be greater. Mean deviation is the average value of the absolute value of the deviation from the mean.

[16] The average price of electricity to industrial customers in 1967 was 0·9c/kWh ('Statistical Yearbook of the Electric Utility Industry for 1971', Edison Electric Institute, New York, p 53). The average price of electricity sold to the primary aluminium industry was 0·34c/kWh, as computed from data for SIC industry category 3334, primary aluminium, on p 23 of 'Fuels and Electric Energy Consumed,' 1967 Census of Manufacturers, US Department of Commerce Document M667(5)-4. We have found a similar difference for 1963.
[17] *Gas Facts*, American Gas Association Inc, Arlington, Virginia, 1970, pp 99-100.
[18] R. Herendeen, 'Affluence and energy demand', presented at the 94th Winter Annual Meeting of the American Society of Mechanical Engineers, Detroit, November, 1973 (Paper 73-WA/ERER-8). Also reprinted in *Mechanical Engineering*, October, 1974. Also available as Document No 102, Center for Advanced Computation, University of Illinois, July, 1973.
[19] R. Bezdek and B. Hannon, 'Energy, manpower and the highway trust fund,' *Science*, vol. 185, p 669, August, 1974.
[20] E. Hirst, 'Food related energy requirements,' *Science 184*, pp 134-138.
[21] E. Hirst and R. Herendeen, 'Total Energy Demand for Automobiles,' Society of Automotive Engineers paper 730065. Presented at the International Automobile Engineering Congress, Detroit, Michigan, January, 1973.
[22] *Merchandising Week*, Vol 104, No 9, 28 February 1972.

Table 1. Energy cost of energy: efficiency of the US economy in delivering energy, 1967

Sector, I/O Number	Efficiency (%)
Coal, 7.00	99·3
Refined petroleum, 31.01	82·8
Electricity, 68.01	26·3
Natural gas, 68.02	90·9

Efficiencies measured fob producer (mine or refinery); an additional energy cost would be associated with marketing. For electricity and gas, producer sells directly to consumer.

common in regulated energy industries. For example, the primary aluminium industry in the USA paid only 38% of the average industrial price for electricity in 1967.[16] Accordingly, our value for the total primary energy intensity of aluminium exceeds Wright's by a factor of 2·5. Most of this difference is due to his constant-price assumption, but some results from the fact that US aluminium imports amounted to about 10% of domestic production, as already mentioned. A similar situation exists for natural gas, which is also regulated in the USA. In 1963, prices to commercial users were more than twice the average industrial rate, and off-peak industrial users on interruptible service received rates much lower than average.[17]

Sectors most affected by preferential prices are primary metals and large manufacturing sectors, where energy prices deviate most from the national average. Since I-O tables are highly disaggregated in those sectors, they contribute heavily to our (equally weighted) computation of the average difference (12%) between our results and Wright's. The true effect of neglecting energy embodied in imports is less than 5% (on the average) for the relatively closed economy of the USA.[13] For a more open economy, imports could not be neglected and as we have seen here, they should never be neglected when we are concerned with the energy intensities of *individual* commodities.

Example applications

Several applications of this method have already been published.[13 18 19 20 21] These usually identify some group as consumers — households, government, specific industries — and energy-cost their purchases. Here we will briefly discuss four applications: the energy cost of energy; the total energy cost of the automobile; the total energy cost of an electric mixer; the total energy import-export balance of the United States.

1. We first obtain the energy cost of energy for 1967. Energy delivered refers to the point of use and 3·80 Joules of primary energy are required to effect delivery of 1 Joule of electricity as electricity, after allowing for losses in mining, generation and transmission. The reciprocals are the energy delivery efficiency of the system, as listed in Table 1. We would emphasise that these are underestimated due to the exclusion of capital flows from the data base; however, from unpublished calculations we feel that the inclusion of capital purchases will decrease these efficiencies by no more than 2%.

2. The total energy cost of the automobile has been computed twice before,[12 21] but we do so again with our latest values for energy intensity. The idea is to determine all auto-related expenditures and apply each to its appropriate energy intensity. Details are in Table 2. We note that our figure for the energy cost to manufacture the average car in 1967, 148×10^9 Joule, is only 11% greater than the result of a detailed process study by Berry and Fels.[5] Final demand expenditures associated with the auto accounted for 19·8% of the nation's energy budget. Only 56% of this was for direct use as fuel.

3. Computing the energy cost of an electric mixer illustrates the relative roles of the energies to manufacture and operate the device. In 1967, an average mixer cost $14 retail.[22] To be exact, we should separate the transportation and trade margins from the manufacturer's price, and apply the appropriate energy intensities.

Table 2. Energy impact of the automobile, 1967[a]

	Expenditure ($10^9$$)	Sector	Energy Intensity[b] (10^3J/$)	Energy ($10^{15}$$)	% of Total
Gasoline					
production	7·28[c]	31·01	—	7400	55·9
refining	—	31·01	(0·208 J/J)	1540	11·6
retail markup	3·50[d]	69·02	37400	131	1·0
Oil					
production & refining	1·09[f]	31·01	—	53[e]	0·4
retail makeup	0·73[f]	69·02	37400	27	0·2
Auto					
manufacture	17·80[g]	59·03	70400	1250	9·4
retail markup	5·93[h]	69·02	37400	222	1·7
Repairs, maintenance, parts	14·31[i] 14·44[i]	75·00	51800	742	5·6
Parking, garaging	14·44[i]	75·00	51800	748	5·6
Tyres					
manufacture	1·11[f]	32·01	104500	116	0·9
retail markup	0·74[f]	69·02	37400	27	0·2
Insurance	11·32[i]	70·04	22000	249	1·9
Taxes (highway const.)	5·94[d]	11·04	123900	735	5·6
Total	84·2			13240	100·00

a The analysis is described in Reference 16. The numbers here differ somewhat since the calculation in Reference 16 was for 1970.
b From I-O calculation results.
c Statistical abstracts of the USA, 1972.
d Petroleum facts and figures, 1971, API.
e 0·51 gallon oil to 100 gallon gasoline. See d above, p 321.
f Purchased cost from c above, Table 896. Retail markup assumed to be 40% of this.
g There were 7·437 x 10^6 cars worth $15 653 x 10^6 (wholesale) manufactured domestically in 1967 (See d above,/p 306). Also 1·021 x 10^6 cars were imported (See c above, Table 892). We assume they had the same unit price of $2104·75 wholesale.
h Retail markup from difference between purchase cost of $2806 (See c above, Table 896) and wholesale price above.
i See c above, Table 896.

For simplicity we assume that all of the margins are allocable to retail trade (I-O sector 69·02). Electric mixers belong in sector 54·04, electric housewares and fans. Then the energy to manufacture and sell the mixer is

$$[0.6 \, \epsilon_{54.04} + 0.4 \, \epsilon_{69.02}] \times 14$$
$$= [0.6 (73\,500) + 0.4 (37\,300)] \, 1.4 \times 10^4 \, J$$
$$= 8.26 \times 10^8 \, J$$

Assuming that the mixer lasts 14 years, with no maintenance or disposal costs, the yearly energy impact of manufacture and sale is 5·9 x 10^7J. Operational energy is obtained by noting that a typical mixer uses about 10 kWh of electricity per year (125 Watts, 13 minutes per day).[23] Taking into account the inefficiency of electricity generation and delivery (Table 1), this is 1·37 x 10^8 J/yr. Then the total yearly energy impact of the mixer is

$$1.37 \times 10^8 + 5.9 \times 10^7 = 1.96 \times 10^8 (0.70 + 0.30) J.$$

70% results from operating the device; 30% from producing it.

4. Figure 3 shows the energy import-export balance for the US in 1967. We recall that there are three ways for energy to enter the economy through imports.

1. Actual energy value of imported energy. For transferred imports, this is $\sum_l P_l$, where the sum is over the energy sectors.
2. Energy penalty associated with energy imports. This is

$$\sum_l (\epsilon_l - 1) P_l$$

3. Energy embodied in imported goods. This is $\sum_{i \neq l} \epsilon_i P_i$

[23] Electric Energy Association, *Annual Energy Requirements of Electric Household Appliances*, EEA 201-73, 1973.

Figure 3: Energy import-export balance for the United States, 1967. Figures expressed as percentages of US energy requirement, which is defined here (following the usual convention) as the sum of domestic energy plus raw imported energy, 66·83 x 10^{18} Joule. Directly allocated imports are estimated and correspond to an energy intensity of 116 x 10^6 J/$. The correct value is probably a bit smaller.

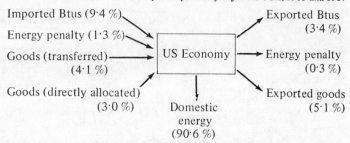

Analogous terms apply for exports. Analogous terms would also apply for directly allocated imports if their energy costs were known.

These terms are evaluated in Figure 3. Imports and exports are expressed as a percentage of the total US energy budget, which we define as the sum of domestic and all imported actual energy.[24]

The energy content of directly allocated imports is estimated at between 2% and 3% of domestic energy use. In terms of actual energy, the USA was a net energy importer (9·4 − 3·4 = 6·0%). In terms of total energy, the US was even more of a net importer [14·8 + approx 3 (for directly allocated imports) − 8·8] = 9·0%.

Updating results

Energy intensities presented here are based on 1967 economic and energy data: ie, on '1967 technology.' How good are they now? Strictly speaking, we don't know. However, since we have results for 1963[23] and 1967 (this report), we can discuss empirical checks of approximate techniques one might use for updating.[25] The problem is data availability. In the USA price indices, GNP, and overall energy use are tabulated annually. We therefore define the following 'options' for updating:

Option 0: Use the old intensities unchanged.
Option 1: Use price indices to correct for inflation (this assumes no technology change).
Option 2: In addition to Option 1, include an overall average change to account for a change in the energy/GNP ratio. Option 2 is what we have used in our past work.

Beyond Option 2 actual technology changes are needed. The most likely place to start is in the technology of energy use: the energy rows of the transactions matrix **X**. We define:

Option 3: Option 1 plus specific changes in the use of energy.

We have applied Options 1, 2 and 3 to the 1963 results in an attempt to update them to 1967 (aggregation to 90-order economy was necessary because of limited data on price indices). For Option 3, we used the actual 1967 energy rows. To compare the projected intensities with the actual 1967 values, we computed

$$\delta_j(\cdot) = \frac{\epsilon_{6j}(1967) - \epsilon_{6,j}(\cdot)}{\epsilon_{6,j}(1967)}$$

where (·) denotes the option used. We treated the δ_j as independent and of equal weight,[26] and calculated mean and mean deviation, thus

[24] Reflecting the flavour of the approach used in this chapter, we should more properly add to this the energy penalty and the energy embodied in imported goods. However, we retain the conventional definition of a country's energy requirements: domestic extraction plus raw energy imports.
[25] R. Herendeen and K. Shiu, 'Comparison of Methods for Projecting Energy Coefficients,' Technical Memo No 47, Center for Advanced Computation, University of Illinois, February 1975.
[26] We use equal weighting because in our work we often use the coefficients singly (eg, how much energy to make steel). Other weightings are appropriate for other purposes (eg, weighting according to final demand expenditures to obtain the energy cost of the whole GNP).

Table 3. Empirical check of techniques to update energy intensity 1963-1967[a]

Option		Mean[b] of δ	Mean deviation[b,c] of δ
0	No update	−0.20	0.20
1	Price indices introduced	−0.16	0.22
2	Option 1 plus overall energy/GNP factor	−0.16	0.22
3	Option 1 and specific updates on energy use in each sector	−0.02	0.07

a Source: Reference 25
b Computed for total primary energy 90 level economy; mean and mean deviation (md) computed for the 85 non-energy sectors only.
c
$$md = \frac{\sum_{I=6}^{90} |\delta(I) - \text{Mean}|}{85}$$

asking how well the options reduce the average 'error' and scatter of the co-efficients. Results are in Table 3.

We see that Option 2, ie, use of price indices and an overall energy/GNP factor was a very poor updating method for 1963-67. In fact, it was very little better than doing nothing at all. On the other hand, Option 3, ie, use of actual 1967 energy-use technology as well as price indices, was quite successful, reducing the average error to 2% and the mean deviation to 7%.

Appendix A

We perform the computations described in Equations (1)-(5) in the section on computational details for a 3-sector economy.

Let the economy be represented by its dollar and energy flows:

$$X = \begin{bmatrix} 10 & 40 & 0 \\ 5 & 5 & 10 \\ 5 & 0 & 5 \end{bmatrix} \begin{matrix} \text{Btu} \\ \text{Btu} \\ \$ \end{matrix}$$

$$\hat{X} = \begin{bmatrix} 50 & 0 & 0 \\ 0 & 50 & 0 \\ 0 & 0 & 30 \end{bmatrix} \begin{matrix} \text{Btu} \\ \text{Btu} \\ \$ \end{matrix}$$

$$\hat{P} = \begin{bmatrix} 0 & 0 & 0 \\ 0 & 10 & 0 \\ 0 & 0 & 10 \end{bmatrix} \begin{matrix} \text{Btu} \\ \text{Btu} \\ \$ \end{matrix}$$

$$Y = \begin{matrix} \text{Btu} \\ \text{Btu} \\ \$ \end{matrix} \begin{bmatrix} 0 \\ 30 \\ 20 \end{bmatrix}$$

$$E_{earth} = (50 \quad 0 \quad 0)$$
$$\quad\quad\quad\quad \text{Btu} \quad \text{Btu} \quad \$$$

For computation it's a bit easier to work with Equation (4).

$$A = X(X-P)^{-1} = \begin{bmatrix} \frac{1}{5} & 1 & 0 \\ \frac{1}{10} & \frac{1}{8} & \frac{1}{2} \\ \frac{1}{10} & 0 & \frac{1}{4} \end{bmatrix}$$

Note that the units of A are:

$$\begin{bmatrix} \frac{\text{Btu}}{\text{Btu}} & \frac{\text{Btu}}{\text{Btu}} & \frac{\text{Btu}}{\$} \\ \frac{\text{Btu}}{\text{Btu}} & \frac{\text{Btu}}{\text{Btu}} & \frac{\text{Btu}}{\$} \\ \frac{\$}{\text{Btu}} & \frac{\$}{\text{Btu}} & \frac{\$}{\$} \end{bmatrix}$$

We first concentrate on crude oil; ie, one type of energy.

$$\epsilon = (100) \begin{bmatrix} \frac{105}{64} & \frac{15}{8} & \frac{5}{4} \\ \frac{5}{16} & \frac{3}{2} & 1 \\ \frac{5}{32} & \frac{1}{4} & \frac{3}{2} \end{bmatrix}$$

$$= (\frac{105}{64} \quad \frac{15}{8} \quad \frac{5}{4})$$
$$\quad \frac{\text{Btu}}{\text{Btu}} \quad \frac{\text{Btu}}{\text{Btu}} \quad \frac{\text{Btu}}{\$}$$

Now we worry about crude and refined as separate energy types. As before, we use Equation (3) for the actual computation:

$$E_{earth} = \begin{bmatrix} 50 & 0 & 0 \\ 0 & 40 & 0 \end{bmatrix}$$

Note that we use the *domestic* refined petroleum figure for E_{earth}.

$$\mathbf{E}_{earth}(\mathbf{X}-\mathbf{P})^{-1} = \begin{bmatrix} 1 & 0 & 0 \\ 0 & 1 & 0 \end{bmatrix}$$

$$\text{and } \epsilon = \begin{bmatrix} \frac{105}{64} & \frac{15}{8} & \frac{5}{4} \\ \frac{5}{16} & \frac{3}{2} & 1 \end{bmatrix}$$

$$\begin{array}{ccc} \text{Btu} & \text{Btu} & \text{Btu} \\ \text{Btu} & \text{Btu} & \$ \end{array}$$

Appendix B
Final demand as the ultimate energy sink

We wish to show that the sum of the domestic and total imported energy (actual fuels and embodied energy) is equal to that energy (actual and embodied) delivered to final demand. We illustrate for the case of one energy type. Rewrite Equation (1):

$$\mathbf{E}_{earth} = \epsilon(\hat{\mathbf{X}}-\hat{\mathbf{P}}-\mathbf{X}) = \sum_{i=1}^{N} \epsilon_i (\hat{\mathbf{X}}-\hat{\mathbf{P}}-\mathbf{X})_{ij}$$

$$\mathbf{E}_{earth} = \sum_{j=1}^{N} (\mathbf{E}_{earth})_j$$

$$= \sum_{j=1}^{N} \sum_{i=1}^{N} \epsilon_i(\hat{\mathbf{X}}-\hat{\mathbf{P}}-\mathbf{X})_{ij}$$

But $\sum_{j=1}^{N} (\hat{\mathbf{X}}-\hat{\mathbf{P}}-\mathbf{X})_{ij} = \mathbf{Y}_i - \mathbf{P}_i$, where $\mathbf{Y}_i$ is the final demand for sector i's output, and

$$\mathbf{E}_{earth} + \sum_{i=1}^{N} \epsilon_i \mathbf{P}_i = \sum_{i=1}^{N} \epsilon_i \mathbf{Y}_i \quad \text{(B-1)}$$

This also shows that the energy intensity of a unit of product i sold to final demand is just ϵ_i. Extension to several kinds of energy follows easily.

This proof has ignored directly allocated imports (P_j refers only to transferred imports). To account for them, we would have to add their embodied energy to both sides of Equation (B-1).

8. The energy cost of goods and services:

an input-output analysis for the Federal Republic of Germany, 1967

Richard V. Denton

Recent studies of energy costs in national economies have tended to concentrate on the UK and the US. This study applies input-output analysis to the economy of the Federal Republic of Germany. It indicates that foreign trade in non-energy products has a great impact on the national energy balance with energy contained in exports being larger than imports by about 25%. It is hoped that the general results will be useful in comparing energy costs in various economies, although explicit comparisons between countries of energy costs per final demand are complicated by the fact that currency exchange rates do not correspond to the rates which should be used when making energy comparisons.

The research was supported by Siftung Volkswagenwerk and conducted in association with the Mesarovic-Pestel World Model Project.

* 1 tSKE = 7 x 10⁶ kilocalories = 8139·5 kWh.
[1] Deutsches Institut für Wirtschaftsforschung (DIW), Quarterly Report 1-74, Dunckner & Humblot Press, Berlin, 1974.
[2] DIW Weekly Report, Berlin, 10 January 1974. Provisional results for 1970, using a different classification than the one in the DIW tables, are available in 'Wirtschaft und statistik', Statistisches Bundesamt Wiesbaden, March 1974.

In calculating sectoral energy cost per final demand, the 1967 input-output tables, at 1967 prices, computed by the German Institute for Economic Research (DIW), have been applied.[1] Although these tables are now eight years old they do represent the most recent complete tables available at present (1975).[2] The DIW tables for 1970 should be available towards the end of 1975.

Data on the energy flows have been taken from *Energy Balances of the Federal Republic of Germany*.[3] Their use is convenient since all the energies have been converted to a common energy unit – tonnes of coal equivalent (tSKE*). Furthermore, the raw data contributed by the member organisations in the 'Working Group for Energy Balances' has been analysed and refined in order to remove inconsistencies due to differences in the data collection bases.

However, economic data compilation need not be on the same basis as energy data collection. As a result, directly matching up the energy data with the sectors in the input-output tables is no trivial task. One way of proceeding is to aggregate the input-output tables so that they correspond to the available energy data. Here the 56 sector DIW input-output table has been aggregated to 26 sectors.

In spite of this aggregation there are still a number of difficulties. The usual problems relating to coupled production processes have to be dealt with, such as when coal is consumed both to produce briquettes, to be sold by the coal sector, and gas, to be delivered to the gas sector. Although the overall energy consumption is accounted for, the actual sectoral energy cost results depend upon the allocations chosen for each sector. Secondly, the sector household-commercial invariably represents a very heterogeneous set of end users, and in fact is a 'catch-all' for energies not accounted for elsewhere in the statistics.

Finally, the import coefficient A_{Mk} of Equation (9) (see pp 86-87 for Equations (1)-(10)) can be obtained readily from the input-output table, but unfortunately this is not the case for the matrix elements α_{mk} which describe the composition of imports going into each sector. At present we have contented ourselves with making plausible estimates of α_{mk} based upon the German economy; the resulting total energy costs e_j^T will be seen to increase considerably from the initial values e_j for some sectors, ie energy costs associated with imports are significant for some sectors. Specific details are available elsewhere.[4]

[3] 'Energiebilanzen der Bundesrepublik Deutschland' Verlags- und Wirtschaftsgesellschaft der Elektrizitätswerke mbH — VWEW, Frankfurt (Main), 1971.
[4] Working paper, ISI, July 1975 (unpublished).

Results

The initial sectoral energy costs, obtained with the use of Equation (5), are displayed as the diagonally lined area in Figure 1. The energy sectors generally have a high energy intensiveness, the coal sector being highest with 31·3 kWh per German mark of 1967. We note that most of the contribution to the energy costs of the coal mining and petroleum refining sectors is due to the fairly large quantities of energy supplied directly to final demands, the terms E_{3y}/Y_3 and E_{5y}/Y_5 of Equation (4) being large.

The apparently low value shown for domestic petroleum and natural gas extraction is a result of 'allocating away' the energy in this sector to the petroleum refining sector, in order to avoid the problem of double-counting mentioned in the Appendix, pp 86-87.

The gas and water sector also has a lower energy intensiveness than the remaining three energy sectors. The gas and water works are

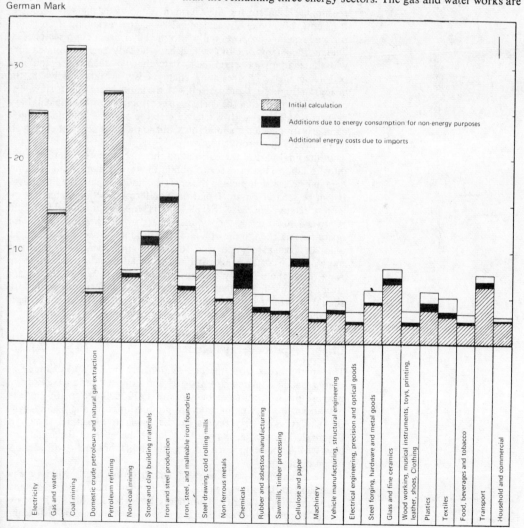

Figure 1: Energy costs per sectoral final demand in kWh per 1967 German Mark

lumped together in the DIW input-output tables and were not disaggregated in the present implementation. If this disaggregation were carried out, with water works properly being included among the other energy user sectors, then the gas sector by itself would presumably have an energy intensiveness comparable to the other energy producing sectors.

Nevertheless, effects related to the energy sectors do not significantly affect the energy costs obtained for the remaining sectors. Of these, iron and steel production has the highest energy intensiveness (14·9 kWh/German mark), while the two sectors, electrical engineering, precision and optical goods and wood working, musical instruments, etc, have the lowest (2·16 kWh/German mark).

Energy costs associated with consumption of energy for non-energy purposes, such as chemical feedstocks, lubrication of machinery, etc, can also be included. Data on the distribution of these energies to the various sectors are not generally available, but a reasonable estimate for the German economy is that 15% goes to transport, 60% goes into chemicals, and the remaining 25% is distributed roughly proportionally into the other production sectors (excluding the energy sectors, which have already been accounted for), with these latter proportions being determined by their consumption of petroleum fuels. The resulting energy costs are also shown in the histogram Figure 1. Comparison with the initial energy costs shows, as one would expect, that the energy costs associated with chemicals have been most affected with a relative increase of 44%. The next largest change is for plastics manufactures, due to the large share of chemicals going into plastics production.

In the final part of the calculation the energy costs e_j were inserted into Equation (10) in order to obtain the total energy costs per final demand e_j^f, which include the energies hidden in imported products. It can be seen in Figure 1 that the non-ferrous metals sector has the largest relative increase (64%). This is due to the inclusion of the relatively high energy costs associated with the foreign mining operations. Although the iron and steel production sector also had additional costs associated with foreign mining operations, relatively more of these costs are allocated to the final demands in other sectors. As a result there is only a relative increase of 10% for the iron and steel sector itself. Whether or not the additional costs due to imports should be counted in an energy analysis depends, as emphasised elsewhere,[5] on the purpose to which the energy analysis is being put.

Import-export energy balance

The energy costs from the last section can be used to examine an energy balance related to imported and exported products. For this purpose the sectoral trade balances of 1970, deflated to 1967 prices, have been used.[6] Multiplication of product imports and exports with the energy cost e_j^f results in the qualitative picture shown in Figure 2. The picture is only 'qualitative' because of the usual problems of associating each of the individual products with an appropriate sector.

The situation is typical of a country not overly rich in raw materials, so that import purchases are greater than exports for most of the primary industries. The energy flows reflect this, with the excess of import over export energies also being shown in the histogram.

[5] See, for example, Chapters 1, 2 and 6.
[6] *Die Wirtschaft 1973*, Deutscher Taschenbuchverlag, 1973.

Goods and services: an input-output analysis for the FRG, 1967

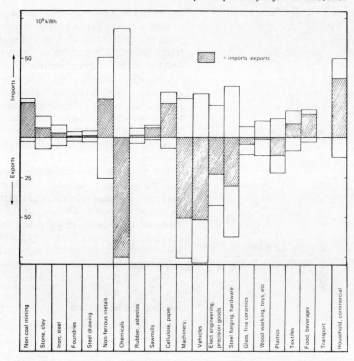

Figure 2: Import-export energy balance of the FRG economy

The chemicals sector is a major exception among the primary sectors, since the significant chemical imports are compensated by an even greater volume of exported finished chemical products.

Investment goods, starting with the machinery sector, show net exports of energy. The transport sector is not shown since it was not included as a separate entity in the data. (Presumably transport costs were included within the individual import and export product prices, and a more accurate treatment would disaggregate this information.) The relatively large imports shown for the last sector, household and commercial, is mainly due to the inclusion of most agricultural products. West Germany has large net imports of agricultural products. The histogram shows a net energy export of about 107×10^9 kWh, energy exports being larger than imports by about 25%. There is no reason to expect an exact balance between the two, since this section has dealt only with the energies allocated to *products*. The actual net energy imports into the energy sectors in 1970, which are not shown in the figure, amounted to 1390×10^9 kWh.

Concluding remarks

No explicit comparisons have been made of the energy costs obtained for other countries. Considerable care is required, since currency exchange rates do not necessarily correspond to the rate which should be used in making energy cost comparisons. For example, the energy costs calculated for household appliances in the US in 1963 were 18·8 kWh/1963$.[5] In terms of the exchange rate in 1963, which was 4 marks = 1 dollar, this would be 4·7 kWh/1963 mark. Since that time there has been a clear recognition that the dollar was overvalued compared with the mark. For the purpose of making a comparison with similar German household appliances of 1963 the value 4·7 would probably be too low a value to use. Conversely, a calculation in real terms (ie inflation corrected), using the present exchange rate, could well be too favourable to the mark.

Related problems arise when one tries to correct for inflation. A complete correction would involve inflators for each sector. However, reasonable average corrections can be obtained by reducing the results given in the present chapter by 1·5, since 1 mark of 1967 = 1·50 marks of 1975 as measured by the price index of the Gross Domestic Product.[7] This of course does not mean that the energy costs are reduced in real terms!

We have also included the imports in the energy cost calculation, and the procedure above glosses over the differences between the domestic inflation rate and the generally higher inflation experienced by other countries. On the one hand, it can be argued that the domestic price levels respond to import price increases, and thus the domestic price index eventually accounts for higher import prices. On the other hand, sufficiently large price increases will tend to induce substitution effects wherever possible; this, however, is outside the realm of the present static calculation.

Appendix: The method

In the standard input-output formalism the interrelationships of an n-sector economy in any given year are described through the equation

$$X_i = \sum_{j=1}^{n} A_{ij} X_j + Y_i \quad (1)$$

where X_i is the total output of sector i, Y_i is the output of sector i sold to the sector final demand, and A_{ij} is the sales of sector i to sector j divided by the total output of sector j. Because of the long time constants involved in the introduction of new technologies, the coefficients A_{ij}, which to a certain extent reflect the underlying technologies of an economy, are assumed to be essentially constant. Thus to obtain A_{ij} it suffices to collect data for a single year in the recent past.

Herendeen[8] introduces energy by writing down the energy flows from each of the energy sectors:

$$E_i = \sum_{k=1}^{n} E_{ik} + E_{iy} \quad (2)$$

where E_i = total energy output in (kWh) of energy sector i,
E_{ik} = energy sales (kWh) for i to k,
E_{iy} = energy (kWh) of type i sold to final demand.
He then writes for each E_{ik}, with the use of Equation (1),

$$E_{ik} = \frac{E_{ik}}{X_k} X_k$$

$$= \frac{E_{ik}}{X_k} \sum_{p=1}^{n} [(1-A)^{-1}]_{kp} Y_p$$

Inserting this into Equation (2), one has

$$E_i = \sum_{k=1}^{n} \sum_{p=1}^{n} \frac{E_{ik}}{X_k} [(1-A)^{-1}]_{kp} Y_p + \frac{E_{iy}}{Y_i} Y_i \quad (3)$$

He defines for convenience

$R_{ik} = E_{ik}/X_k$ and
$S_{ik} = \begin{cases} E_{iy}/Y_i & i = k = \text{one of the energy sectors} \\ 0, & \text{otherwise.} \end{cases}$

Written in vector notation his result is equivalently

$$E = [R(1-A)^{-1} + S] Y = \epsilon Y \quad (4)$$

which defines the energy matrix ϵ. Once the data have been collected and the energy matrix ϵ has been calculated according to Equation (3), the energy flows from each of the energy sectors for a given year are determined if the final demands are specified.

In the present paper we are interested in calculating the energy costs e_j for the economy of the Federal Republic of Germany, as obtained by summing the

---

[7] 'Statistische Beihefte in den Monatsberichten der Deutschen Bundesbank,' seasonally adjusted, June 1975.
[8] R.A. Herendeen, 'Use of Input-Output Analysis to Determine the Energy Cost of Goods and Services,' in *Energy: Demand, Conservation and Institutional Problems*, edited by M.S. Macrakis, The MIT Press, 1974.

Goods and services: an input-output analysis for the FRG, 1967

elements ϵ_{ij} of the energy matrix over all energy sectors i:

$$e_j = \sum_i \epsilon_{ij}$$

$$= \sum_{i,p} R_{ip} [(1-A)^{-1}]_{pj} + S_{jk} \quad (5)$$

An element e_j gives the total consumption of energy from all energy sectors necessary for the economy to deliver a unit monetary value of a product from sector j to final demand Y_j.

The terms of E_{ij} in Equation (3), where i=k, can be taken to be the consumption of energy within the energy sector i; this internal consumption is then allocated to the 'true' users, final demands, Y_j in Equation (3). Note: a direct application of Equation (3) would account for some energies twice, and therefore the sum of Equation (3) over all energy sectors i would add up to more energy than is actually consumed in the economy in any given year. For example, the coal sector provides (primary) energy for the generation of electricity in the electricity sector, and in turn the electricity sector supplies (secondary) energy back to the coal sector. Adding both of these flows would involve double-counting of energies.

This problem can be avoided by following Herendeen's suggestion of initially allocating the relevant primary energies to the secondary energy producing sectors. In the example above any coal used in the electricity producing sector is to be included within the electricity sector itself, and the flow $E_{coal \to electricity}$ is set equal to zero.

After calculation of the energy matrix using the method, the primary energies which have been included in the secondary sectors must be allocated back to their original sectors by introducing a new allocation matrix C (for details see Herendeen). However, we are only interested here in calculating $e_j = \Sigma \epsilon_{ij}$, and it turns out that this final reallocation is not necessary. The fact that the total energy must be conserved in the reallocation places certain mathematical constraints on the elements of C, and as a result the matrix C drops out in the sum over i in the calculation of e_j.

So far this procedure allocates all energies used in the domestic market to final demands. For some purposes it is reasonable to include, in addition, the energy costs associated with products which are imported into the domestic market. (It would be misleading to state a low energy cost associated with some final demand item if all energy intensive operations in its manufacture have been transferred outside the country.) In the following we describe a procedure which allocates these additional energies associated with imports to their respective final demands.

Total imports X_M are distributed to the various manufacturing sectors X_k and to final demand for imports Y_{Mf} according to

$$X_M = \sum_k A_{Mk} X_k + Y_{Mf}$$

where A_{Mk} is the sales of imports to sector k divided by total output of sector k (for energy sectors 'sales of imports' *excludes* actual energy imports, since these have already been counted in Equation (3); it refers only to the 'non-energy' products). In turn, an added energy flow is to be associated with these imports:

$$E_M = \sum_k \tilde{e}_k A_{Mk} X_k + \tilde{e}_{Mf} Y_{Mf}$$

$$= \sum_{k,p} \tilde{e}_k A_{Mk} [(1-A)^{-1}]_{kp} Y_p$$

$$+ \tilde{e}_{Mf} Y_{Mf} \quad (6)$$

Here the energy cost factor $\tilde{e}_k$ is the energy per unit monetary value of the products which are imported into sector k, and $\tilde{e}_{Mf}$ is a similar factor corresponding to final demand for imports.

The previous results, Equations (4) and (5), are then combined with Equation (6) to give the total energy cost (including imports) of a sector p, here denoted by e_p^T;

$$e_p^T = e_p + \sum_k \tilde{e}_k A_{Mk} [(1-A)^{-1}]_{kp} \quad (7)$$

To calculate e_p^T we require the energy cost factors $\tilde{e}_k$, which have not yet been given. Obviously, if $\tilde{e}_k$ had to be determined for all countries which contribute to imports, the calculation would be somewhat time-consuming. A way to avoid this unpleasantness is to assume that the sectoral energy costs of other countries is the same as the corresponding sector of the Federal Republic. Then the simplest approximation to $\tilde{e}_k$ would be a single weighted average of the e_j values calculated from Equations (4) and (5), with the weighting related to the sectoral composition of the imports, and then to set all the $\tilde{e}_k$ equal to this result.

Averages however can be misleading and, at the expense of adding a distribution matrix α_{pk} the energy cost factors $\tilde{e}_k$ can be expressed in a better approximation as

$$\tilde{e}_k = \sum e_p^T \alpha_{pk} \quad (8)$$

The matrix element α_{pk} gives the fractional share of the imports going into sector k which have the energy cost of sector p of the domestic market. Since the total of the fractional shares for a branch k must add up to 1, one has $\sum_p \alpha_{pk} = 1$. Also, we note again that $\tilde{e}_k$ for an energy sector k refers to non-energy imports into k, since energy imports are already counted in Equation (4).

The use of $\tilde{e}_p^T$ instead of just e_p from Equation (5) can be interpreted fairly simply. In order to calculate the total energy cost of the sectoral final demands, including the energy hidden in imports, one must treat the energy cost of the imports themselves with the same convention. (Another country would calculate its *own* total energy costs including *its* imports, say e_p^{T1}, and a closed solution is obtained by our assumption

$$e_p^{T1} = e_p^T$$

Combination of Equations (7) and (8) results in the following expression for e_p^T:

$$e_p^T = e_p + \sum_m e_m^T G_{mp} \quad (9)$$

where

$$G_{mp} = \sum_m \alpha_{mk} A_{Mk} [(1-A)^{-1}]_{kp}$$

The solution in vector notation is

$$e^T = e \cdot (1-G)^{-1} \quad (10)$$

In the above **e** is the sectoral energy cost excluding imports as obtained in Equation (5). Multiplication with the inverse matrix $(1-G)^{-1}$ provides the additional energy costs hidden in imports which must be added to the sectoral final demands, to give the total energy costs.

87

9. Energy analysis of nuclear power stations

Peter F. Chapman

Responding to the controversy stirred by his initial papers analysing nuclear power, the author has refined the original analysis, correcting its omissions, and aims to document the factual basis on which energy analysis of nuclear power is based. He concludes that differences in opinion on this application of energy analysis are due not simply to differences in data or the adoption of different analytical conventions, but rest upon fundamentally conflicting views of consumer behaviour and how energy is used.

This chapter, describing an energy analysis of nuclear power stations including a discussion of the conventions and methods used, is largely based on the research report published in 1974 by Chapman and Mortimer[1] (referred to as ERG 005). This report was rapidly supplemented by papers by Price,[2] Leach[3] and Wright and Syrrett.[4]

The resulting controversy has shown a number of weaknesses in both the initial study and many of the criticisms directed against it. Much of the controversy rests on fundamental misunderstandings of both the methods and applications of energy analysis, exacerbated by a failure of many critics to read either the energy analysis literature or ERG 005 and by the failure of the authors of ERG 005 to incorporate explicit behavioural assumptions or demand forecasts in their conclusions.

The aims of this chapter are to document the factual basis on which the energy analysis is based and to correct the omissions in the initial report. It is hoped that this will demonstrate that the differences of opinion concerning the application of energy analysis to nuclear systems are not due to differences in data, nor due to arbitrary differences in analytical conventions but are due to fundamentally different views concerning consumer behaviour and the future. Although this chapter cannot resolve the controversy it is hoped that it will point to the controversial factors and thereby remove many of the misunderstandings.

Before describing the analysis it is important to establish the aims of the study since only then can the methods and conventions used be judged. The initial study was aimed at evaluating the fossil fuel input to different nuclear reactor systems and investigating the role of uranium ore grade on the viability of thermal reactor systems. In the course of this investigation the oil-crisis prompted many industrial nations to announce crash nuclear programmes to overcome their dependence on OPEC oil. These announcements prompted the extension of the project into the dynamic analysis of nuclear programmes. Here the initial study evaluated nuclear power as a direct substitute for oil, ie, as a primary fuel. Since then the study has been enlarged to consider nuclear systems as sources of electricity, introducing questions of the relative utility of different fuels. The study is continuing, with the emphasis now focused on the energy required to obtain uranium from deposits of different ore grades.

---

[1] P.F. Chapman and N.D. Mortimer: *Energy Inputs and Outputs for Nuclear Power Stations* Open Univ Rsch Rpt ERG 005, Dec 1974.
[2] J. Price: *Dynamic Energy Analysis and Nuclear Power* Friends of the Earth, Dec. 1974. (Available FOE 9 Poland St, London, W1)
[3] G. Leach: *Nuclear Energy Balances Re-examined*, Int Inst for Env and Dev, London 1974.
[4] J. Wright and J. Syrett: 'Energy analysis of nuclear power *N. Scientist* 65 (No 931) p 66, Jan 1975.

It should be emphasised that the initial study and the developments described here do not rest in any way on an 'energy theory of value'. Energy analysis is a descriptive, not prescriptive, technique. It may provide an input to policy evaluation, just as environmental impact statements or statements concerning levels of employment may provide inputs. However to be directly useful to policy makers the results of energy analyses have to be combined with behavioural assumptions and value judgements. One of the benefits of the controversy is that it has made explicit a number of previously implicit assumptions and value judgements.

Energy inputs and outputs

The methods of energy analysis and the need to establish certain analytical conventions have been described previously.[5] There are a number of important conventions in this analysis concerning the energy requirement (e.r.) attributed to imports, to uranium and to plutonium. The e.r. attributed to imported commodities is set equal to the energy required for their production in the exporting country. Although this reflects the depletion of world fuel stocks due to the production of the commodity, it has been argued that, since the UK has not expended energy in producing imports, they should be given a zero e.r. This latter argument ignores the fact that the UK does have to expend energy producing commodities which it exports in order to purchase the imports. Furthermore it is likely that if an import is especially energy intensive then this may be reflected in a relatively high price, requiring the export of either a large volume of low energy intensity goods or a similarly energy intensive commodity.

Since the purpose of the initial study was to evaluate the fossil-fuel input required to produce nuclear power it was decided to attribute no energy content to uranium in the ground. Thus the only e.r. attributed to uranium is the energy which has to be expended in mining, milling and processing ore to produce uranium. This convention is not meant to imply that uranium has no energy content in a thermodynamic sense. It is simply a useful convention for the purpose of the analysis. (It also avoids questions such as evaluating the thermodynamic potential of uranium and spent fuel rods.) In order to be consistent with this the nuclear power station is not given any credit for its plutonium production, ie, both uranium and plutonium are considered as *materials*, not as fuels. (If the power station were to be given a credit for its plutonium then it would be consistent to debit the station with the uranium which it has fissioned.)

Throughout the analysis of a nuclear station all the inputs and outputs are normalised to a nominal installed capacity of 1000 MW (net electrical). The fuels consumed in the production processes have been given the energy requirements described in a previous chapter[6] except that the electrical inputs have been kept separate. It has so far proved impossible to examine all the production processes involved in detail. The major inputs to the construction of the power station, for example the buildings and electrical equipment, have been given energy requirements based on the energy intensities (ie the ratios of total energy required to value of output) deduced from input-output studies[7] and analyses of the census of production.

In principle this is an unsatisfactory procedure since the inputs to nuclear systems are likely to be uncharacteristic products of the

[5] Chapter 1: 'Energy costs: a review of methods'.
[6] Chapter 2: 'The energy cost of fuels'.
[7] D.J. Wright: 'The Natural Resource Requirements of Commodities' *Applied Economics* 7 p 31 1975.

Table 1. The financial costs and energy requirements for the capital equipment.

Reactor Type	Electrical		Buildings		Nuclear Steam System		Total Energy Requirements	
	£m	TJ(th)	£m	TJ(th)	£m	TJ(th)	TJ(e) + TJ(th)	g.e.r. TJ(th)
MAGNOX	52	6178	73	7358	116	12528	1738 + 19112	26064
SGHWR	52	6178	31	3125	67	7236	1102 + 12131	16539
PWR	52	6178	30	3024	50	5400	973 + 10710	14602
AGR	52	6178	30	3024	89	9612	1254 + 13798	18814
CANDU	52	6178	31	3125	67	7236	1102 + 12131	16539
HTR	52	6178	30	3024	60	6480	1045 + 11502	15682

sectors documented in the input-output tables. However there are grounds for believing that provided a product has a large vector of inputs, ie requires inputs from many other sectors of the economy, then the average energy intensity derived from the input-output table is fairly reliable. If a product requires only one or two inputs, such as the manufacture of a particular chemical, then it may have a real energy intensity (as evaluated by process analysis) very uncharacteristic of the input-output table sector. This is particularly true of the chemicals used in mining and processing uranium, uranium itself and the production of heavy water, all of which have to be analysed by process analysis.

In order to calculate the energy requirement of the capital equipment inputs to the power station, the electrical equipment, buildings and the nuclear steam system, we need to know their financial costs and the energy intensities of the appropriate sectors of the economy. Overall costs have been given by the CEGB[8] and these have been broken down by comparison with the costs of coal-fired stations.[9] The breakdown for a number of reactors has been confirmed by TNPG[10] and is shown in Table 1 together with the total energy requirements.

The energy intensities deduced from the input-output tables are for 1968 and these have to be adjusted for inflation when applied to the 1973 cost estimates. The division of the total energy requirement into thermal and electrical components is based on the proportions of electricity consumption documented in the Census of Production.[11]

Core requirements

In order to evaluate the energy requirement of the initial core of the reactor it is necessary to know the quantity of uranium, its enrichment, the grade of uranium ore mined and the energy inputs to each of the processes shown in Figure 1. The quantities of material shown in Figure 1 are those required to produce the 160 te of 2·1% enriched uranium needed for the initial core of a 1000 MW SGHWR from a 'typical' US uranium mine. The ore grade, stripping ratio and inputs to a 'typical' uranium mine were evaluated using the details of four US mines given by Everett.[12]

Weighting the mines in proportion to their estimated reserves gives an average ore grade of 0·31%, an average stripping ratio of 24:1 and an average energy requirement of 1210 MJ/ton ore mined. There are several different milling and extraction processes used in uranium mines.[13] The average energy required for milling and beneficiation is 1224 MJ/ton ore, comprising 396 MJ/ton of direct fuel (largely electricity), 792 MJ/ton due to the consumption of chemicals, water

[8] CEGB evidence to Select Committee on Science and Technology *The Choice of Reactor System.* HMSO (ISBN 0 10 276574 X), Appendix 6 p 192.
[9] Searby, P.J: *Atom* 178 p 185 (Aug 1971).
[10] Personal communication from The Nuclear Power Group, Warrington: 1974.
[11] *Report on the Census of Production 1968* HMSO 1971.
[12] F.D. Everett: *Mining Practices at Four Uranium Properties in the Gas Mills, Wyoming* US Bureau of Mines Circ No 8151, 1963.
[13] J.W. Clegg and D.D. Foley: *Uranium Ore Processing*, Addison-Wesley, 1958.

Energy analysis of nuclear power stations

Table 2. Natural uranium and SWU requirements per tonne of enriched uranium (based on 0·25% tails assay)

Fuel enrichment (%)	Tonnes natural U per tonne fuel	SWU per kg fuel
1·6	2·94	1·25
1·9	3·59	1·75
2·1	4·02	2·05
2·45	4·78	2·70
2·6	5·11	3·0
2·8	5·54	3·4
3·0	5·98	3·8
3·3	6·63	4·45
3·7	7·50	5·30
6·5	13·59	11·28
10·0	21·20	19·08

[14] T.M. Pigford, M.J. Keaton and B.J. Mann: *Fuel Cycles for Electrical Power Generation* Teknekron Inc, Rpt No EEED 101 1973 (2118 Milvia St, California 94704 USA).

[15] J.T. Roberts: 'Uranium enrichment: supply, demand and costs' *Int. Atom. Energy Bull.* 15(5). Oct 1973 p 14.

[16] C.F. Barnaby, 'The gas centrifuge project' *Sci. J.* 5A(2) 1969 (Aug) p 54.

etc, 22 MJ/ton due to machinery and parts and about 14 MJ/ton due to the plant establishment. The inputs for the conversion of U_3O_8 to UF_6 have been taken from Teknekron's study of electricity fuel cycles.[14]

The inputs and outputs of the enrichment plant are characterised by the enrichment of the final fuel rod and the tails assay, i.e. the concentration of U^{235} remaining in the discarded fraction. In order to achieve a higher enrichment or a lower tails assay the enrichment plant has to perform more 'separative work', which requires a larger energy input. Throughout this analysis the tails assay is taken as being 0·25%. This fixes both the quantity of natural uranium required to produce a tonne of fuel enriched to a given percentage and the number of 'separative work units' (SWU's) required per kg of fuel produced. These are documented in Table 2. The electrical input is taken to be 2·42 MWh/SWU[14,15] (8·71 GJ(e)/SWU). To this must be added a fraction of the capital energy requirement, estimated to be 2·4 £/SWU[16], equivalent to an e.r. of 0·35 GJ(th)/SWU.

The final stage of the fuel preparation process is the conversion of the enriched UF_6 to uranium dioxide and the assembly of fuel rods. The energy requirements for this stage have also been taken from the Teknekron study.[14] Table 3 shows the total e.r.'s for reactor fuel preparation broken down into the e.r. for producing uranium, its enrichment and its fabrication.

To utilise this data in calculating the e.r. of the initial core of a reactor we need to know the quantity of uranium and its enrichment, which is not easy to obtain. Much of the published literature on reactors contains data which nuclear engineers and the UKAEA have

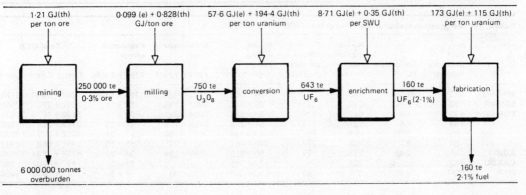

Figure 1: The processes involved in producing reactor fuel. The quantities of materials are those required to produce the initial core of an SGHWR reactor.

testified as being false, or at least uncharacteristic. In particular there are wide variations in the quantities of uranium (per MW capacity) and the levels of enrichment.

In order to make realistic comparisons of different reactors we have used the data for specific power stations, taking only self-consistent data from the Nuclear Power Index[17] and the Directory of Power Reactors.[18] The Magnox data refers to 'Oldbury A', the SGHWR data to the design described by Moore,[19] the PWR's are (i) Haddam Neck; (ii) Maine Yankee; (iii) Jos M. Farley and (iv) Shearon Harris, the AGR refers to Hunterston B, the CANDU data is for Pickering and the HTR data is from a TNPG design.[10] The combination of the reactor parameters with the energy inputs shown in Table 3 is summarised in Table 4. The last column gives the total e.r. of the initial core assuming that the electricity comes from a fossil-fired station.

To complete the energy inputs to nuclear power stations we need to know the inventories of heavy water and the energy required for its production. Data from the Heavy Water Division of Atomic Energy Canada gives an e.r. of 650 MWh(e) + 6000 MWh(th) per tonne D_2O. The CANDU reactor has an inventory of 0.7 te/MW(e) and the SGHWR has a planned inventory of about 0.25 te/MW(e). Table 5 gives the total energy requirement of the nine different reactors analysed.

Although the grand totals are very similar the division between plant, core and other costs is very different. It is particularly noticeable that the CANDU system requires much less uranium than the other reactors. There have not been many other analyses of all the

[17] 'Index of Nuclear Reactors' *Nucl Eng Intl* April 1972.
[18] *Directory of Power Reactors.* IAEA Vienna, 1971.
[19] J. Moore, N. Bradley and I.T. Rowlands: 'The SGHWR' *Atom 195* (Jan 1973) p 7.

Table 3. Energy requirements for uranium fuel production.

	GJ(e) + GJ(th)	Total (assuming 25% electricity efficiency)
To produce one ton natural uranium from 0.3% ore (stripping ration 24:1) including conversion to UF_6	96.5 + 995.5	(1381)
To perform one tonne SWU of enrichment	8710 + 350	(35900)
To convert and fabricate one ton of uranium fuel	173 + 115	(807)

Table 4. The energy requirements for the initial core of 1000 MW(e) stations.

	Fuel Inventory (tonnes)	Enrichment (%)	Natural uranium required (tonnes)	SWU per core (te SWU)	Mining, milling Energy TJ(e)+TJ(th)	Enrichment energy TJ(e) + TJ(th)	Fabrication energy TJ(e) + TJ(th)	Total GER TJ(e) + TJ(th)	TJ(th)
MAGNOX	973	natural	973	—	94 + 968	—	168 + 112	262 + 1080	2128
SGHWR	160	2.1	643	328	62 + 640	2857 + 115	28 + 18	2947 + 773	12561
PWR (i)	130	3.3	862	578	83 + 858	5034 + 202	22 + 15	5139 + 1075	21631
(ii)	104	2.6	531	312	51 + 529	2718 + 109	18 + 12	2787 + 650	11798
(iii)	97	3.35	654	456	63 + 651	3972 + 160	17 + 11	4052 + 822	17030
(iv)	87	2.7	463	278	45 + 461	2421 + 97	15 + 10	2481 + 568	10492
AGR	195	2.45	932	527	90 + 928	4590 + 184	34 + 22	4714 + 1134	19990
CANDU	182	natural	182	—	17.6 + 181	—	31 + 21	49 + 202	396
HTR	22.7	6.5	308	256	30 + 307	2230 + 90	4 + 3	2264 + 400	9456

Table 5. Total energy required for 1000 MW(e) power stations

	Capital Equipment			Heavy Water			Initial Core			Total		
	TJ(e)	+ TJ(th)	%	TJ(e)	+ TJ(th)	%	TJ(e)	+ TJ(th)	%	TJ(e)	+ TJ(th)	TJ(th)
MAGNOX	1738	+ 19112	92				262	+ 1080	8	2000	+ 20192	28192
SGHWR	1102	+ 12131	45	585	+ 5400	21	2947	+ 773	34	4634	+ 18304	36840
PWR (i)	973	+ 10710	40				5139	+ 1075	60	6112	+ 11785	36233
(ii)	973	+ 10710	55				2787	+ 650	45	3760	+ 11360	26400
(iii)	973	+ 10710	46				4052	+ 822	54	5025	+ 11532	31632
(IV)	973	+ 10710	58				2481	+ 568	42	3454	+ 11278	25094
AGR	1254	+ 13798	48				4714	+ 1134	52	5968	+ 14932	38804
CANDU	1102	+ 12131	43	1638	+ 15120	56	49	+ 202	1	2789	+ 27453	38609
HTR	1045	+ 11502	62				2264	+ 400	38	3309	+ 11902	25138

inputs to nuclear power stations. The study by Hill and Walford[20] checked the above calculation and used the same data. A Westinghouse study[21] gives a 30% higher energy requirement for mining, a 10% smaller e.r. for enrichment, much higher (by factor of 10) e.r. for fabrication and the same e.r. for the plant to give a total energy input 4% higher than that given above. Studies still in progress in Sweden[22] and France[23] have produced very similar estimates of the energy input to the plant (excluding the initial core) based on very disaggregated input/output studies. The Bechtel corporation[24] have claimed to have produced a much smaller result but have consistently refused to publish any details of their calculations. Ignoring for the moment differences in conventions there appears to be good agreement on the data and there are not any grounds for believing that the + 15% error estimated in ERG 005 is unrealistic.

Similar remarks apply to the calculation of the net outputs of the 1000 MW reactors, although here there are much larger differences in assumptions. One significant difference between various authors has been the assumed load-factor. ERG 005 used a 62% life-averaged load factor based on CEGB conventions. This assumes that at the beginning of its life the station will be operated at a load-factor of 70-80%. In practice the UK Magnox reactors have only achieved this after being down-rated by 15% and the average load factors of PWR's operating in the USA in 1973 was slightly less than 60%. It has recently become apparent that rather than argue about which is the 'correct' load factor to use in the analysis it would be more profitable to examine the sensitivity of any particular conclusion to the load-factor chosen.

Refuelling and distribution losses

The second conventional problem in calculating the net output of the nuclear power station concerns the subtraction of the energy used in refuelling the reactor. The problem arises because some of the energy used in producing fuel rods is a thermal requirement and some electrical. It is obvious that the electrical component should be subtracted directly from the electrical output of the station. In ERG 005 the thermal component was similarly directly subtracted. This procedure is repeated here since, for the high-grade ores considered, it is not particularly significant. This will be considered in more detail later.

In addition to allowing for the station load-factor and the energy requirements for refuelling it is also necessary to subtract electricity

[20] K. Hill and F. Walford: 'Nuclear Aspects of Energy Accounting' presented at conf. 'Understanding Energy Systems' London, 1975.
[21] R.J. Creagan: *Net Output of Energy from Nuclear Sources* Oct 1974. Available from R & D Planning: Westinghouse Electric Corp, 700 Braddock Ave, East Pittsburgh, Penn
[22] I. Stöhl: Verbal report given to IFIAS workshop. Stockholm 1975.
[23] Charpentier: Verbal report at IFIAS 1975, Stockholm.
[24] W.K. Davis: 'A nuclear plant pays back its energy investment in 2-3 months'. Report in *Nucleonics Weekly* 16(1). March 1975, p 7.

lost in distribution, electricity used by the electricity industry and the energy required to make up the heavy-water inventories. It should be noted that the quantity of electricity used by the electricity industry itself is estimated on the basis of past usage and assumes that future use will be in proportion to the total installed capacity. This assumption may be unreasonable for very rapid growth rates. The net power and energy outputs of the reactors examined are documented in Table 6.

It should be emphasised that the energy outputs documented in Table 6 are directly proportional to the assumed station lifetime, in this case 25 years. There is general agreement about the net *power* output of nuclear stations, but the use of different station lifetimes can lead to very different estimates of the total energy output. This means that the energy ratios for the reactors, ie the ratio of energy out to energy in, is sensitive to the lifetime assumption. In contrast, the power ratio for a reactor, the ratio of output power to input power, is sensitive to the assumed construction time since the input energy is well defined. To remove the relatively arbitrary assumptions of construction time and lifetime it is useful to consider the ratio of input energy/output power which can be interpreted as the 'payback time' for the reactor. These three parameters for each of the reactors are shown in Table 7.

Dynamic analysis of building programmes

The 'payback times' shown in Table 7 are in the range 1·5 to 2·5 years and indicate the length of time it takes the power station to produce an amount of energy equal to E_{in} once the station has started producing an output. However it takes about five years to construct a station and about one further year to bring the reactor up to criticality and check out the installation and so the station produces an energy profit between 7·5 and 8·5 years after the beginning of the construction stage. This is significant because in this period a number of other power stations may be commissioned, and these will increase the energy input before the first station has produced an energy profit indicating that at the beginning of a rapidly growing nuclear power building programme a substantial energy deficit may arise. This will only occur when the doubling time of the growth in the number of power stations is significantly less than the overall payback time for one reactor (including the construction and commissioning times).

Table 6. The outputs of 1000 MW(e) stations (assuming 62% L F; 7·5% distribution loss; 3·75% own use)

	Net Output Outside Elec. Ind. (MW)	Refuelling Power (MW)	Power for D_2O Make Up (MW)	Net Output (MW)	Energy Out in 25 Years TJ(e)
MAGNOX	525·25	8·8	—	541·45	426,880
SGHWR	"	23·4	3·45	423·4	412,650
PWR (i)	"	49·2	—	501	395,024
(ii)	"	27·8	—	522·4	411,858
(iii)	"	32·5	—	517·8	409,993
(iv)	"	24·7	—	525·5	414,378
AGR	"	31·8	—	518·45	408,708
CANDU	"	3·2	5·25	541·8	427,248
HTR	"	42·6	—	507·7	400,248

Table 7. Reactor parameters. (Uranium from 0·3% ores)

	Energy Ratio (assuming 25y lifetime)	Power Ratio (assuming 5y construction)	Payback Time (E_{in}/P_{out}) (years)
MAGNOX	15·1 ∓ 3	3·02 ∓ 0·5	1·65 ∓ 0·25
SGHWR	11·2 ∓ 2	2·24 ∓ 0·4	2·23 ∓ 0·3
PWR (i)	10·2 ∓ 2	2·18 ∓ 0·4	2·29 ∓ 0·3
(ii)	15·6 ∓ 3	3·12 ∓ 0·5	1·60 ∓ 0·25
(iii)	12·9 ∓ 2	2·58 ∓ 0·4	1·93 ∓ 0·3
(iv)	16·5 ∓ 3	3·30 ∓ 0·5	1·15 ∓ 0·25
AGR	10·5 ∓ 2	2·10 ∓ 0·4	2·38 ∓ 0·3
CANDU	11·1 ∓ 2	2·22 ∓ 0·4	2·25 ∓ 0·3
HTR	15·8 ∓ 3	3·16 ∓ 0·5	1·58 ∓ 0·25

For example if the number of nuclear stations were doubling every two years then if there was one station finished in year 1 we would require one further station to be finished in year 3, two more to be finished in year 5 and four more in year 7. Thus by year 7 the programme would have 8 stations finished. Since the construction and commissioning time is six years, this means that in year 1 it is necessary to have 7 stations under construction in addition to the one station finished. Whilst the growth continues exponentially then the number of stations under construction will always be about seven times the number of stations (see the Appendix, p. 101).

If the seven stations under construction require a larger energy input than can be provided by the output of the one station finished then this building schedule will require a net energy input for as long as the growth continues. In order to use the data on inputs and outputs in the previous section to evaluate the energy balance of this schedule we have to establish a convention for comparing the mix of electrical and thermal inputs with the electrical output. There are a number of conventions available to resolve this problem, all of which depend upon certain behavioural assumptions. In order to clarify the discussion of these conventions let us consider a programme for building SGHWR's, the programme having a doubling time of 2 years. When one 1000 MW station is completed it will produce an annual output of 16505 TJ(e). Assuming that the energy inputs to the 7 stations under construction are spread out uniformly over the 5 year construction period the annual energy requirement will be 6487 TJ(e) *plus* 25625 TJ(th).

Convention A: This starts by observing that one nuclear power station does not significantly alter the overall efficiency of generating electricity in the UK. Furthermore the nuclear electricity is evenly distributed to all consumers (via the national grid). Hence the electrical input to the nuclear construction industries 6487 TJ(e) requires 25948 TJ(th) of fuel, which combined with the 25625 TJ(th) input gives a total input (of fossil fuel) equal to 51573 TJ(th). Assuming that the nuclear output is used as a fuel (and not exclusively for work) its output, 16506 TJ(e), is subtracted directly to given an energy *deficit* of 35067 TJ(th) per year per GW completed. While the building schedule grows exponentially the fuel deficit gets larger and larger as more stations are completed.

Convention B: The starting point for this convention is that the electrical inputs to the nuclear construction industries are assumed to come from nuclear power stations. Subtracting the 6487 TJ(e) input from the 16506 TJ(e) output leaves a net output of electricity of 10019 TJ(e). Now it is also assumed that this electricity is purchased by consumers who presently use fossil fuels for performing work. Since electricity is about three times more useful than a fossil fuel for performing work the consumption of 10019 TJ(e) *saves* 30057 TJ(th) of fossil fuel. (If the consumers substitute nuclear electricity for fossil fuel then this is actually the fossil fuel released for other consumption.) This fossil fuel saving is larger than the required fuel input 25625 TJ(th), so according to this convention the two year doubling programme actually shows an energy profit of 4432 TJ(th) per year per GW completed.

Convention C: This is the convention which I feel is most realistic in the sense that it appears to be closest to the way in which fuels are actually consumed. The above conventions produced apparently opposite answers because they made extreme assumptions about consumer behaviour. Convention A is unrealistic because some nuclear operations (particularly enrichment) are quite likely to get 'special' electricity, possibly from a nuclear station. Convention B is unrealistic because it assumes that all future users of electricity will only use it for work previously performed by fossil fuels. In fact a large proportion of electricity is already used for heating, perhaps as much as 25%. Convention B would be a good approximation if for every 1GW of nuclear capacity finished 1GW of fossil-fired capacity were closed down (thereby guaranteeing the assumed substitution). Without this assumption it would seem more reasonable to subtract the electrical input to the nuclear construction industries from the nuclear output (leaving 10019 TJ(e) net output as before) and then subtract the thermal inputs 25625 TJ(th) from this net output. This gives a net energy deficit of 15606 TJ(th) per year per GW completed which is almost exactly mid-way between the conclusions reached by conventions A and B.

If convention C is applied to a programme where the number of reactors is doubled every 4 years, which gives a ratio of stations under construction to stations finished close to 2, then it is found that there is a net energy profit of 10825 TJ(e) per year per GW completed (roughly half the gross output of the finished reactor). Although this makes it appear that this slower programme is preferable it should be borne in mind that the slower programme takes longer to reach any total capacity. Figure 2 shows how the two programmes compare assuming a starting capacity of 5GW and a final capacity of 80 GW.

The period when the fast programme is a net consumer of fuel is almost exactly balanced by the period (from year 8 to year 16) when it produces more output than the slower programme. The choice between these programmes depends upon whether the deficit can be 'afforded' and when the output is 'needed'. Again this emphasises that while such an energy analysis may help by providing information on the consequences of certain policies, energy analysis alone cannot tell you which of these programmes is 'preferable'.

There are a number of other features of this type of dynamic analysis which should be mentioned. Firstly the above results describe the consequences of certain building schedules and the only implicit comparison is between adopting a particular schedule or none at all. This is quite a different type of analysis to that performed by Leach[3]

Figure 2: Comparison of two nuclear programmes which have a maximum capacity of 80GW and a starting capacity of 5GW. One programme has a doubling time of 2 years, the other a doubling time of 4 years.

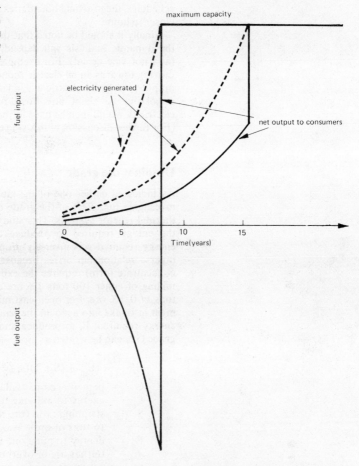

and Hill and Walford[20] who both compared the nuclear schedule with one based on fossil-fired power stations. As it happens both these analyses chose growth rates which avoided the very serious deficit identified above (Leach used a doubling time of 4 years and Hill and Walford a doubling time of about 6 years). It is not surprising therefore that these authors observed a substantial saving of fossil fuels in the nuclear programme compared to the fossil-fired programme. This is not to say that even more fuel could be saved by not pursuing either programme.*

Perhaps the most important comment on this type of analysis is that it is almost certainly irrelevant in the real world. The point is simply that if a nuclear building programme can be shown to run into a net energy flow problem then its net cash flow problem will be about 10-20 times worse, since the purchase of fuels represent only 10% of the total costs involved. Not only will the cash-flow problems be horrendous but so too will be the problems of the construction industries, electrical engineering trades etc, all of whom would have to increase their outputs very quickly — and worse, come to a grinding halt at the end of the growth phase. Although this type of energy analysis may impose a theoretical boundary on feasible building

*For example had Hill and Walford assumed a demand for space heating, and not for electricity, then their conclusions would have been quite different.

schedules, the practical boundaries will be found a lot sooner and a lot closer to home.

Finally it should be noted that the most appropriate convention for the dynamic analysis will depend, to a degree, on policy decisions (and not vice versa!). For example if a country adopted a policy of moving towards an all-electric economy then all the energy inputs and outputs would be electrical. Under such circumstances it would be impossible to claim any 'fossil-fuel' saving by substitution since electricity would then be used for both work and heating applications. (For further discussion of this see reference 29.)

Uranium ore grade

As indicated earlier one of the motives of the original study was to investigate the effect of the grade of uranium ore on the viability of thermal reactor systems. This study was suggested by an analysis of the energy required to produce copper[25] which showed that the energy required was inversely proportional to the grade or ore. The inverse relationship arises because to produce 1 ton of copper (in concentrate form) requires the expenditure of energy in mining and milling of either 100 tons 1% ore or 500 tons of 0·2% ore or 1000 tons of 0·1% ore. For open-cast mining the total energy requirement must also take into account the removal of overburden. In general the energy required, E, to produce a ton of any material from an ore of grade G% can be written as

$$E = \frac{100}{G}(E_m + (S+1)E_d) + E_f$$

G = percentage ore grade
E_m = energy to mill one tonne of ore
S = stripping ratio (the ratio of tons of overburden to tons of ore)
E_d = energy to mine one tonne of material (either ore or overburden)
E_f = energy to convert benificated ore to required material.

It should be noted that all the energies listed above refer to gross energy requirements, including the energy required for chemicals, transport, machinery parts etc. This model suggests that there is some grade of uranium at which the energy required to produce one ton of uranium will equal the energy produced from that ton of uranium in a thermal reactor. This will be a grade much lower than the 0·3% ores described previously and for much lower grade ores we can ignore the energy required to convert benificated ore to a finished material (E_f) so that the total energy required is inversely proportional to ore grade.

Table 6 showed that a 1000 MW(e) SGHWR station would produce an output of 412650 TJ(e) in 25 years. This total takes into account the energy for refuelling the reactor. If we do not subtract the refuelling energy requirement then the net output rises to 521·8 MW and the energy output in 25 years is 430977 TJ(e). Subtracting the energy required to build the station, to produce the heavy water inventory and to enrich the initial core and subsequent fuel gives a net yield of 403076 TJ(e). This output is produced from 3743 tonnes of natural uranium, 643 tonnes in the initial core (Table 4) and 126

[25] P.F. Chapman: The energy required to produce copper and aluminium from primary sources'. *Metals and Materials* 8(2) p 107, February 1974.

Energy analysis of nuclear power stations

Table 8. Energy requirements for mining and milling Chatanooga shale (0.007% U_3O_8)

	GJ(e) + GJ(th) per ton U_3O_8	
Mining		
Electricity	2531	1·4
Water		1·4
Tunnel machine spares		285·5
Explosives		350·1
Drill bits		756·2
Diesel fuel		444·1
Roof bolts		77·0
Maintenance		74·2
Supplies		52·2
Capital equipment		245·9
Milling		
Electricity	5422	
Water		23·4
Coal (for steam raising)		12082
Mill rods		116·3
N-B-A		112·7
Kerosene		63·4
Sodium Carbonate		101·2
Ammonia		16·9
Sulphuric Acid		2066·8
Flocculent		79·2
Mill materials		37·1
Capital equipment		664·3
Total	7953 +	17649·9 (= 25603)

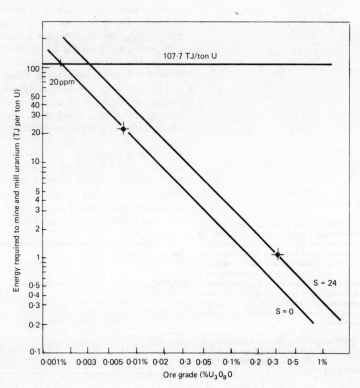

Figure 3: The predicted cut-off grade of uranium ores based on the model and data described in the text.

tonnes per year for refuelling. Thus the SGHWR produces a net yield of 107·7 TJ(e) per tonne of natural uranium. This is the upper bound on the energy requirement for mining and milling uranium, indicated by the horizontal line on Figure 3.

To date the research programme underway has only obtained satisfactory energy requirements for the 0·3% ores (described earlier) and for the extraction of uranium from Chatanooga Shales.[26] The inputs to the Chatanooga shale process are converted to gross energy requirements in Table 8. (This includes several corrections to the estimates made in ERG 005.[1]) Now, if the model relating energy requirement to ore grade and stripping ratio is correct, these two data points can be used to estimate the uranium ore grade corresponding to the maximum energy requirement calculated above. On a log-log plot the model described earlier predicts a straight line relationship between energy requirement (E) and ore grade (G) with a slope of -1 and an intercept which depends upon the value of the stripping ratio (S). These predictions are shown on Figure 3 corresponding to the two ore grades with different stripping ratios. (Note that the energy requirements for uranium do not include any electricity conversion factors.) This suggests that the lower bound on uranium ore grade is about 20 ppm.

Clearly a lot more data on different uranium mines, with different ore grades, different rock hardness etc is needed to substantiate this estimate. However this estimate is compatible with the values of E_d deduced from the copper study[25] and the value of E_m calculated for uranium benification.[13] It is also important to realise that at this cut-off ore grade the activity is producing no net energy output, ie it is an outer bound. In practice it is unlikely that a system would prove viable in simple economic terms without a yield of 5-10 times the energy input in the mine and a more practical limit would be about 100 ppm. (An SGHWR fuelled from such ores would have a power ratio of about 0·4, an energy ratio of about 2 and a 'payback time', as defined in Table 7, equal to 12·5 years.)

This practical limit is clearly not much better than an 'order of magnitude' estimate since it depends upon the efficiency of all the mining and milling machinery, the particular properties of a given deposit, the net yield from the reactor and other technological factors. For example it may happen that a lower grade deposit in a remote region with abundant hydro-electric power may prove to be worth exploiting; in effect using the uranium as a means of transporting the hydro-power to other parts of the world. But what the analysis has demonstrated is that for uranium deposits with a grade close to or less than 100 ppm there is a very good case for examining the viability both in energy and financial terms.*

* It should be emphasised that the price mechanism would show whether a particular ore grade was worth exploiting, but only when the mine was needed (ie no higher grade deposits being available). The energy analysis is able to anticipate a *possible* non-viability, but is not able to state definitely which deposits will prove uneconomic. Anticipation is important due to the long lifetimes of nuclear stations.

[26] Bieniewski, C.L. et al: *Availability of uranium at various prices from resources in the US.* US Bureau of Mines. Inf Circ 8501, 1971.

Table 9. Estimated uranium reserves

Ore Cost ($ per lb U_3O_8)	Concentration (ppm)	Reasonably Assured (m. tonnes)	Awaiting Discovery (m. tonnes)
Less than 10	700 – 2500	1·0	1·0
10 – 15	450 – 1600	0·7	0·7
15 – 30	250 – 800	1·0	1·0
30 – 50	140 – 500	0·5	1·0
		3·2	3·7

Appendix: Mathematical analysis of construction programmes

Let the number of stations started in year t be given by n (t) where

$$n_i(t) = a \exp(at)$$

The cumulative number of starts up to year T is given by n(T) where

$$n(T) = \int_0^T n_i(t)\, dt = \exp(aT) - 1$$

This function is zero when $T = 0$ which is convenient. The number finished at the time T is given by

$$n_f(T) = \int_6^{T-6} n_i(t)\, dt = \exp[a(t-6)] - 1$$

and the number under construction, $n_c(T)$, is given by

$$n_c(T) = \int_{T-5}^{T} n_i(t)\, dt$$

$$= \exp(aT)\,[1 - \exp(-5a)]$$

Provided that we are well into the programme then $n_f(T) \gg 1$, so that $\exp[a(T-6)] \gg 1$, hence

$$n_f(T) \simeq \exp[a(T-6)]$$

Then the ratio, $n_c(T)/n_f(T)$ is given by

$$\frac{n_c(T)}{n_f(T)} \simeq \frac{\exp(aT)\,[1 - \exp(-5a)]}{\exp(aT) \cdot \exp(-6a)}$$

$$= \frac{1 - \exp(-5a)}{\exp(-6a)}$$

This shows that provided $n_f(T) \gg 1$ then the ratio of stations under construction to those finished is independent of T. Evaluation of this ratio for doubling times of 2 years and 4 years gives 6·57 and 1·64 respectively.

---

[27] R.D. Vaughan: 'Uranium Conservation and the Role of the GCFBR.' Paper presented to Brit Nucl Eng Soc London, January 1975.
[28] *Uranium: Resources, Production and Demand.* OECD. August 1973.
[29] P.F. Chapman: 'The energy analysis of nuclear power' *New Scientist* Vol 68 No 971, Oct 1975.

If the effective cut-off grade for uranium deposits is about 100 ppm then this could have serious implications for the maximum thermal capacity which can be installed. Vaughan[27] has indicated that there may be 6·9m tonnes of uranium available at a price less than 50$/lb U_3O_8 (1970 $). This estimate, together with the approximate ore grades is shown in Table 9, and emphasises that most of this uranium is still 'awaiting discovery'. According to the OECD[28] the total uranium stock is to be divided three ways, between the USA, Europe and the rest (sic!). Of the 2·3m tonnes for Europe the UK could expect to get a fifth, ie 460 000 tonnes of uranium. Since 1000 MW (or 1GW) for 25 years requires 3743 tonnes, this uranium resource estimate for the UK is equivalent to 3072 GWy. It is interesting that current UKAEA plans would lead to a thermal reactor capacity of 4075 GWy (with a peak installed capacity of 104 GW).

Conclusions

The controversy over the energy analysis of nuclear power stations has not arisen as a result of differences in data. The energy inputs and outputs described on pp. 89ff. have been largely confirmed by other published analyses. The conventional differences arising in the comparison of electrical and thermal energy inputs and outputs described on pp. 90ff. do not arise as a result of arbitrary choices of convention but reflect fundamentally different assumptions concerning consumer behaviour. In order to further resolve these differences it would be necessary to examine the ways in which electricity is presently used in society and the ways in which it would be used in the future. These questions are not answerable by the methods of energy analysis.

The energy analysis of extracting uranium from low grade deposits could provide a more direct input to policy formulation. At the present time there is insufficient data available to evaluate the cut-off grade with any accuracy. Such data as is available suggests that the bound may correspond to a grade of 20 ppm U_3O_8 and if this is confirmed by further studies then it suggests that, unless special facilities are available, ores of a grade 100 ppm U_3O_8 or less may not be economically viable. In its turn such a limit of viable uranium deposits implies an upper bound for the total thermal reactor capacity. This type of conclusion, which rests upon the identification of a boundary condition on a well-developed technology (mining), is less prone to misinterpretation and is likely to be the area in which energy analysis proves most useful.

10. Nuclear power and oil imports:

A look at the energy balance

J.H. Hollomon, B. Raz and R. Treitel

There are still a great many arguments over the net energy benefits – if any – of replacing fossil fuel power stations with nuclear. Here the authors have used a simple model in an attempt to calculate the net energy costs of a number of possible strategies. They include that nuclear power can be effective in saving fossil fuels but that the saving is critically dependant both on the rate of reactor construction and the amount of energy needed to construct each reactor. For any given energy construction input there is an optimum rate of construction; programmes faster or slower than the optimum will result in less net energy output.

The growth rate of total electricity consumption of the US has been very stable over the last 20 years and the installed capacity similarly has grown exponentially at about the same rate of 7%/year (Figure 1a. The installed capacity is an aggregate quantity that conceals the structural changes which have occurred in the electrical sector, namely, the significant shift of fossil base load power plants from coal to oil, and more recently, the rise of nuclear power and the increase of orders for new coal fired plants.

To assess the impact of such a rapid growth on the energy balance of the nation, we have used a very simple model describing, on the one hand, the flows of power invested, mainly in the construction of new plants but also in related industries, and on the other hand, the flow of power produced by the new power plants entering operation. The difference is the net power output, which is a rough measure of the incremental useful energy available each year for the remainder of the economy.

We have simulated the net electrical output of the subsystem of new electrical power plants using a model based on the historically proven exponential growth of the installed capacity (Figure 1a). The main input parameters in this calculation are: the yearly energy output of a standard power plant, the yearly energy input in the construction of a standard power plant, the lead time for construction and the lifetime of the plant. (See the Appendix, p.108, for the details of our calculations.)

The choice of an identical set of parameters for coal and nuclear power plants is an approximation too crude to be useful. The coal fired plants in themselves constitute a system which is much more diversified than the nuclear reactors in size, level of technology, and flexibility of operation. In order to avoid complication of the calculations at this preliminary stage, we have confined ourselves to a system of identical nuclear power plants. We keep in mind that this is only one part of the total system of power production, but, as will become apparent later, its study provides information on the general energetic features of a process of growth.

In Figure 2, we display the net power output of a system of identical nuclear power plants at the 21st year of its growth as a function of various doubling times. (The doubling time is inversely related to the rate of growth. For a more exact definition, see

Figure 1a: Past trends of electric utility sector: installed generating capacity and total electricity consumption historical growths.

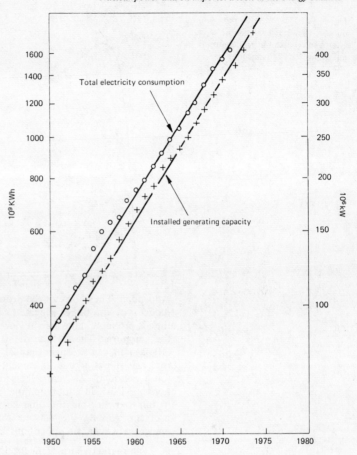

Figure 1b: A fit of an exponential growth law characterized by a doubling time of $D \simeq 4.4$ years to a forecast on future demand made by Edison Electric Institute.

Fig 1b table. US generating capacity in service at year's end (10^3 MW)

Year	Total	Nuclear	% Nuclear (rounded)
1974	406	36	9
1980	636	111	17
1985	924	276	30
1990	1344	508	38
1995	1953	818	41

Source: Edison Electric Institute, 1974[2]

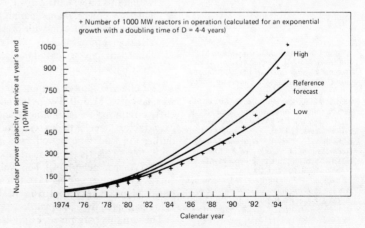

103

Figure 2: The 1995 net energy output of a nuclear reactor system internalising the energy investment in the expansion of the system.

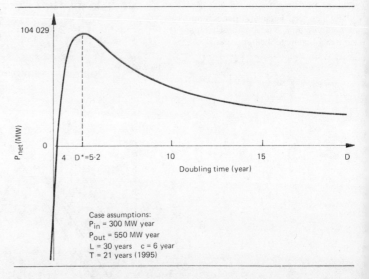

Case assumptions:
P_{in} = 300 MW year
P_{out} = 550 MW year
L = 30 years   c = 6 year
T = 21 years (1995)

Appendix.) The choice of the proper values of the parameters for the model proved a rather delicate task. The value of 6 years for the lead time is probably quite safe in view of the minimum of 5 years cited in the literature.[1] The lifetime of 30 years affects the calculation very little as long as it is longer than 21 years. The Edison Electric Institute forecast[2] shows a good agreement with an exponential law of growth for a period of 21 years. This determines the choice of our time horizon: 1995. In a more remote future, there is bound to be a ceiling to this growth,[3] but the time horizon 1995 allows us to avoid this problem. Among the reasons for which the development rate of nuclear power has an upper limit are:

- The market share of electricity in the total energy production is expected to be limited to about 60% in 1995. This limit is determined by the opportunities to substitute installations that are consuming energy in a non-electrical form (such as domestic heating), and that are not likely to be completely decommissioned within this time span.
- Nuclear power plants cannot be used directly for peak load electricity production with present technology and costs. The cost effectiveness of nuclear plants is, and presumably will remain, limited mainly to base-load generation, about 70% of the total electricity production.

The effective power output (P_{out}) of a nominal 1000MW reactor was taken to be 550MW, after allowing for process inputs like the energy required for enrichment and fuel fabrication, line losses and load factors (see Appendix). The yearly static power input (P_{in}), during the lead time period, was considered constant and equal to 300MW. (This is equivalent to a total energy investment of 1800MW-year per plant.) This value, estimated on the basis of the energy content of capital investment in the construction of a nuclear power plant, is very approximate. It does concur, however, with the value estimated by Chapman.[4]

The variable parameters in Figure 2 are the net output of the whole

---

[1] Bupp I.C., Derian J.C., Donsimoni M.P., Treitel R. (1974), *Trends in Light Water Reactor Capital Costs in the US: Causes and Consequences.* MIT Center for Policy Alternatives, Cambridge, Mass.
[2] Edison Electrical Institute, *Report of the subcommittee on uranium enrichment facilities to the EEI Policy Committee on atomic power* (1974).
[3] Leach G. (1974) *Nuclear Energy Balances in a World with Ceilings.* International Institute for Environment and Development, London, UK.
[4] Chapman P.F., Mortimer N.D., *Energy inputs and outputs for nuclear power stations*, Research Report, ERG005, Open University, Milton Keynes, UK, (Dec 1974).

system of nuclear power plants (P_{net}) and the time required to double the number of existing plants (D). The main objective of the decision maker is assumed to be the control of the net output, P_{net}.

The most interesting feature of the curve P_{net} (D) is the fact that there is a maximum of P_{net} which is positive and which indicates that the choice of the doubling time D of the system may be critical. At D values that are smaller than the maximizing P_{net}, there is a sharp decline in the net output which may become considerably negative at doubling times smaller than D=4 years. If maximizing P_{net} is not considered critical, there is a choice between two values of the doubling time D which produce the same P_{net} (P_{net} being smaller than the obtainable maximum). A choice of a larger D might lower the strain on the economy, while the shorter D will result in 'accumulating' generating capacity of electricity at a period when the economy can afford to spend energy. This is, in a way, equivalent to the storage of energy in a form that will make it available on a long-term basis. In Figure 2, the maximum net output is obtained at D=5·2 years. It might be mentioned that the installed nuclear capacity of the United Kingdom is expected to double each 4·3 years.[5] This value would place the British system in a region of the curve where the net output is very sensitive to small variations of the chosen doubling time.

The large uncertainty about the value of P_{in} has led us to calculate a family of curves of P_{net} as a function of D for different plausible values of P_{in}. (Figure 3.) It becomes apparent that the smaller the value of P_{in}, the higher the maximum of P_{net}, and it is obtained at shorter doubling times. If we compare the curve of P_{net} (D) computed for P_{in} = 300MW in Figure 3 to that in Figure 2 (where P_{in} was also taken as 300MW), we notice that the curve is flattened and its maximum shifts from D = 5·2 years to D = 6·8 years. The shape difference between the two curves is a result of the different time horizons (1995 and 1985, respectively) for which they have been calculated. We conclude that one has to plan differently the rate of expansion of a nuclear power plant system for different time horizons in order to meet the objectives of a specified growth strategy (maximum oil displacement, accelerated storage, etc.).

The curve in Figure 2 has been drawn assuming an exponential growth for the nuclear reactor population. The justification for this assumption is historical in nature, as are many of the prevailing forecasts for the needs for energy in the next 50 years. One of the main purposes of this chapter is to demonstrate to the decision maker that the possible strategies in expanding the power producing sector are in fact quite numerous.

Different exponential rate values can be applied in anticipation of, or in order to influence future developments in the availability of oil. In the same framework of thought non exponential growth rates, say, linear, can be adopted. Also the expansion of the nuclear power plant system might stop or level off because of the development of rival technologies, or because of conservation measures. This does not seem likely in the near future, but one has to bear in mind that if exponential growth is stopped, the net power output of the system will increase considerably.

A word of caution should be heard concerning absolute values derived from the model. Most of the parameters' values are only estimates or even 'best guesses', so that it might be presumptuous to

[5] Price J. (1974), *Dynamic Energy Analysis and Nuclear Power*. Friends of the Earth, London.

Figure 3: The 1985 net energy output of a growing nuclear reactor system for different values of yearly energy investment.

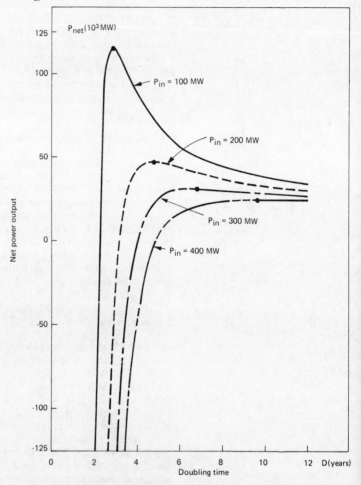

attach an immediate consequence to the numerical results. Yet, the very steep behaviour of the curve in Figure 2 at D values smaller than the maximum, indicates an unstable situation in which a relatively small error in the choice of the pace of execution of a country's nuclear programme may produce significantly different results in the energy balance.

Of the numerous ingredients of the energy issue there are four which have direct relevance to our problem:

- The thermodynamically efficient use of energy resources.
- The economically efficient management of electrical power production.
- The social costs of electrical power production.
- The political and economic aspects of independence of foreign oil. (Freedom from another oil embargo threat.)

The preceding analysis is mainly relevant to the first aspect. Some people however ignore the dynamics of energy inputs in the growing system. For them:

Each 1000 megawatt nuclear plant licensed can produce the energy equivalent of 10 million barrels of oil per year. There can be no denying the fact that nuclear plants can relieve some of the pressures on the fuel oil situation, while at the same time helping this nation achieve President Ford's goals of reducing U.S. oil imports by one million barrels per day by the end of 1975 and reducing the number of oil-fired generating plants by 1980.[6]

Our results point to the conclusion that, even if a nuclear programme has a zero net power output, it still displaces a non-negligible amount of oil. But this will certainly be less than 10 million barrels of oil per year per power plant. To illustrate this take two possible cases: (1) A nuclear programme designed to replace existing oil fired plants; (2) A nuclear programme compared with an oil fired plant programme in order to satisfy future demand.

For the first case we have derived a crude estimate of the amount of oil which can be displaced by alternative nuclear programmes by comparing a single nuclear power plant and an oil-fired plant of identical power output. If we ignore the static energy input in the nuclear plant, it displaces 2·5 times the oil equivalent of its energy output, because the efficiency of conversion of oil into electrical power is at best 40%.

In the dynamic case the total output of a system of reactors is smaller than the sum of the outputs of the single units: $P_{net} < N_{op}P_{out}$ (where N_{op} is the number of operating nuclear plants). However, even in the case of $P_{net} = 0$, where the energy input needed to sustain the growth of the nuclear system is equal to its entire output, at least 1·5 times the oil equivalents of its power output are displaced. In Table 1 we present the amount of oil displaced by a nuclear power plant system for two values of the doubling time and for different values of the power input.

In the second case, dynamic conditions apply to both nuclear and oil fired plants. Energy inputs in construction have to be taken into account in the oil fired plants as well. This should increase the oil displacement performance of the nuclear reactor plants. The extent of this increase depends on the dynamic parameters of the oil fired system. It is clear, however, that the displacement performance of the first case will be the lower limit of the second case.

From Table 1 it can be seen that a nuclear power system could be a useful means for affecting oil demand, although the effect is substantially less than that obtained by simple summation of the individual outputs of each reactor. The amount of oil displaced is sensitive to the choice of an appropriate doubling time for the system.

[6] Murtzing L.M., Speech to the Butcher and Singer, SM Stroller Corporation, 4th Annual Nuclear Conference, quoted in *Information from ERDA*, vol 16, No 4 (Jan 22, 1974).

Table 1. The amount of oil displaced in the year 1980 by a system of nuclear power plants for different doubling times and different values of input power*

P_{in} MW	Total oil displacement (million bbl/day)	
	D = 4·2 yr.	D = 5·2 yr.
100	1·8	1·7
150	1·7	1·7
200	1·5	1·6
250	1·4	1·5
300	1·3	1·4
350	1·2	1·3
400	1·1	1·2
450	0·95	1·2

*This table has been computed for the case where nuclear plants replace existing oil-fired plants.

In order to choose a doubling time that will optimise the level of oil displacement it is necessary to reduce the uncertainty of our knowledge of P_{in}.

Our presentation of the choices is limited to a nuclear-oil comparison. This is in fact an oversimplification of the actual range of choices. We still think that in the case analysed in this chapter there are enough elements of general interest to merit the discussion. It would be useful to extend this type of analysis to energy economy, mixing various primary sources such as coal, gas and hydroelectric power as well as nuclear power and oil.

Appendix: Preliminary model calculation

The model is based on the following conditions:

- A system of identical power plants.
- Throughout the period considered all the plants have the same lead time C and lifetime L. No technological improvement is taken into account.
- The growth of the installed capacity is assumed to be exponential:

$$N_{op}(t) = N^o e^{at}$$

where a is the rate of growth and N^o is a scaling factor: the number of plants in operation when $t = 0$. The exponential doubling time D

$$= \frac{\ln 2}{a}$$

- The unit output power (P_{out}) is identical for all plants.
- The input energy E_{in} is uniformly distributed over the lead time so that the input power is constant (P_{in})

$$E_{in} = C \times P_{in}$$

The number of plants under construction at time t

$$N_c(t) = \int_{t}^{t+c} \frac{\partial N_{op}}{\partial t} dt + \int_{t-L}^{t-L+c} \frac{\partial N_{op}}{\partial t} dt$$

The first integral represents the plants entering operation between time t and t+c. The second integral represents the plants decommissioned between time t and t+c which have to be replaced.

$$N_c(t) = N^o e^{at} (e^{ac} - 1)(1 + e^{aL})$$

The net total power output of the system

$$P_{net}(t) = P_{out} N_{op}(t) - P_{in} N_c(t)$$

$$P_{net}(t) = N^o e^{at} [P_{out} - P_{in}(e^{ac} - 1)(1 + e^{-aL})]$$

The model has been applied for a case of a nuclear power system: Each unit has a nominal capacity of 1000 MWe.

The scale factor (N^0) is determined by the historical installed capacity at time $t = 0$ (1974) which was 36 GWe.

$$N^0 = 36.$$

We assume that a 1000MWe nuclear plant has a capacity factor of ·7 (presently less than ·6 on an average basis) and consumption on site plus distribution losses represents about 10 to 15% of the gross output (80MWyr). The electricity requirement for the enrichment of a yearly load is estimated to be about 3 to 5% of the electrical output (for LWR and present enrichment technology) (35 MWyr). Thus an optimistic value for P_{out} is:

$P_{out} = 1000 \times ·7 - 80 - 30$
$= 590$ MW year.

Evaluations of P_{in} are much more difficult. To our knowledge no serious assessment has been attempted, but using values derived from studies on other energy intensive industrial products, present best guesses are between a (probably too low) 50 MWyr and a high 400 MWyr (for a 6-year lead time). The computer run of Figure 2 has been performed, assuming $P_{in} = 300$ MWyr and $P_{out} = 550$ MWyr.

The authors are very grateful to Mr A.B. Lovins for a stimulating discussion about the issue of net energy considerations, and to Dr M.A. Sirbu, Jr. for his keen interest and useful criticism.

11. Energy analysis of a power generating system*

K.M. Hill, F.J. Walford and R.S. Atherton

The application of energy analysis to nuclear power has provoked considerable controversy about both methodology and conclusions. This chapter looks at the results of a UKAEA developed computer programme applied to a variety of nuclear scenarios. It concludes that energy analysis can add to understanding of energy problems but that the results can be misleading if applied to a single resource like fossil fuel. Any meaningful investment in nuclear power is not likely to repay its investment energy for at least four years after the first station is commissioned.

In an appended section (pp 121-125) the energy costs of the constituent parts of a nuclear reactor are detailed, with special reference to the newly adopted British SGHWR.

As with all forms of measurement and accounting, energy analysis is usually carried out with a specific viewpoint in mind and as a result particular indicators, efficiencies, ratios etc are developed. The analytic conventions are similarly influenced, since the methods of treating data depend very much on what they are going to be used for. Much of the current interest derives from concern about the external aspects of energy systems, such as resource utilisation and environmental impacts, and so naturally the conventions adopted have tended to show up issues of this type.

In the engineering sense energy analysis is just a further example of the realisation of the importance of the systems approach to the use of technology as a servant to society. One must be concerned with what happens outside the factory fence which is where an energy budget differs from the chemical engineer's 'heat balance'. If, therefore, this paper has an air of 'deja vu' it is because as engineers we see energy analysis as an extension of existing methodology rather than as a new one.

The financial accountant at first sight has a considerable advantage over the energy analyst in that he is dealing in uniquely valued and defined units whereas for the energy accountant there are problems of equivalence of thermal energy, free energy, electricity and so on, although all may be expressed in the same units. In fact, the financial accountant also gets into progressively more difficult problems of equivalence the more comprehensive his accounts become, until in economics, we find the economists borrowing the concepts of thermodynamics to help them out. It is not surprising then that the energy accountant or 'ergonomicist' finds himself looking hopefully at economic concepts. He is merely travelling backwards along the economist's path.

These different approaches have led to different types of energy analysis dependent on the end use of the analysis – Table 1. Currently fossil fuel accounting occupies the centre of the stage for three reasons.

- Concern with fossil fuel resources
- Convenience of conversion of economic data into proxy fuel data
- Concern with impacts of energy policies

Apart from the apoplexy felt by the nuclear reactor designer on finding his precious brain child described as a converter of sooty hydrocarbons, there are serious reasons for doubting this approach.

*This chapter is reprinted by kind permission of the United Kingdom Atomic Energy Authority.

- Fossil fuels are not the only fuels or even the only ones in potentially short supply.
- The alternative chemical uses of fossil fuels do not constitute a serious resource problem when this type of use is reduced to strictly chemical terms, eg, it is hydrogen rather than natural gas that is used to produce fertilisers; hydrocarbons form a relatively small resource of hydrogen and hydrogen does not require carbon as intermediate in its production.
- Accounting in terms of a commodity becomes misleading if one is considering situations in which the commodity itself is not used or is being substituted for.
- Commodity accounting is not the best way to illuminate the problems of conservation or extension of resources which should be tackled from a systems viewpoint.

Typical of this work is that of Chapman and Mortimer[1] who have analysed reactor performance. They develop the inputs of fossil fuels to extraction, processing and fuel preparation stages as well as to the capital investments involved and compare them with the outputs of electrical power. A feature of their method is to develop an index of performance — the energy ratio, E_r defined as:

$$E_r = \frac{\text{Output} - \text{running costs}}{\text{Capital input}}$$

This is thus in the classical return on investment form.

The calculation of an energy ratio as a means of indicating the performance of a particular reactor requires the reduction and averaging of a large amount of systems data over the reactor life, which they then re-expand to develop an over simplified system. This unnecessary condensing and expanding of the data obscures, as one might expect, a great deal of useful systems information.

One may quibble with their numbers at a detailed technical level and in the case of utilising low grade uranium ore these quibbles can make a large difference to the comparison of reactor types, but their analysis demonstrates clearly the distribution of energy inputs over

[1] P.F. Chapman & N.D. Mortimer 'Energy inputs & outputs for nuclear power stations' ERG005 Open University 1974, Dec.

Table 1 — Forms of Energy Analysis

Form of energy analysis	Insight desired	Users	Basis of conventions*
Fossil fuel accounting i.e. analysis in units of fossil fuel consumed	Resources depletion, environmental impacts	Environmentalists, resource groups etc.	Result expressed in terms of fossil fuel utilisation. All energy utilisation/conversion systems regarded as devices for burning fossil fuels. No intrinsic value attached to energy.
Fuel demand/production analysis	Utilisation patterns, changes in demand patterns, production capacity required.	Fuel and energy suppliers, planners.	Economic & technical conventions
Energy systems analysis (including process analysis)	Effectiveness of utilisation of energy/free energy	Energy technologists	Standard thermodynamic conventions
Thrift analysis	Wastage of energy	Civil servants	—
Energy input/output analysis	Energy content of goods, use of energy in the economy	Service to other users, policy analysts.	Economic conventions associated with matrix techniques
Energy flow analysis (as in this paper)	Comparison of energy systems using different fuels	Systems analysts	Separation of all forms of energy

*All present forms of Energy Analysis ignore the energy content of labour.

the cycle. In what they do with these results, however, we have severe reservations.

For power stations they define the energy ratio as:

$$E_r = \frac{\text{Net power output} - (\text{thermal \& power}) \text{ inputs of refuelling}}{\text{Total thermal inputs of capital}}$$

The thermal inputs of fuel are used directly without conversion to equivalent power. These inputs are mainly used in practice as power but supplied as fuel, eg for diesel engines. In principle such power could be supplied directly by electricity whether from fossil or nuclear sources. It follows that for consistency the debit should be the power obtained from the fossil fuel used and not the power required to generate the required thermal input.

Correction of the energy ratio for this defect is difficult since the power equivalent of the thermal inputs of fuel, particularly for the mining and ore-processing costs which dominate the low grade fuel case, would be difficult to determine. The error introduced by this misconception is small where the energy inputs into fuel production concern high grade uranium ores but is large with low grade ores. The ratios calculated by Chapman & Mortimer will need to be changed. However we regard the use of energy ratios to be unsatisfactory and prefer to derive the energy flows, as described later.

For all these shortcomings the Chapman Mortimer approach of treating fossil fuel accounting as an investment problem has brought out the valuable concept of looking at the energy required to set up programmes over time as a criterion for judging energy policies, particularly in a situation where, in the short to medium term, one is tied to fossil fuels, both to supply current consumption and investment requirements. The fact that, in their early analysis, they obtain results that *appear* to condemn nuclear power is not a function so much of nuclear power but of the scenarios they investigated. This has been pointed out by Leach[2] who considers both the effects of limited demands for power as against continued exponential growth and the cumulative effect of substitution of fossil fuel by nuclear. This is entirely consistent with the evidence in our previous paper[3] and in this paper.

Energy flow analysis

In a system as complex as the UK electricity industry there are difficulties in predicting response to technological change. It was, therefore, thought worthwhile to simulate different growth and technology scenarios and to show the energy flows directly.

To do this, use was made of DISCOUNT G[4] computer programme developed by Central Technical Services Branch of the UKAEA. The programme calculates the stocks and flows (which may be expressed in money, material or energy terms) arising from a given electricity-generating programme. One can build into the programme constraints relating to upper limits on the rate of building of any particular types of station, limits on their load factors and life and a merit order for the introduction of station types to meet load variation. Given a pattern of demand, the programme then calculates the time of introduction of additional stations necessary to meet the demand and their operating load factors. There was an additional built-in constraint that fast reactors were not introduced into the building programme until

[2] Gerald Leach *'Nuclear energy balances in a world with ceilings'*. Int Inst for Environment and Development, London, Dec 1974.
[3] K.M. Hill & F.J. Walford 'Nuclear aspects of energy accounting' Inst Fuels & OR Society Conference – Understanding energy systems 15–16 April 1975, London.
[4] C.E. Iliffe *'DISCOUNT G: A digital computer ...'* UKAEA Risley, June 1973.

plutonium stocks had reached a level adequate to fuel them.

We have derived the energy flows for construction of plants and provision of fuels in place of the costs for which the programme was designed, using a zero rate of discount. The DISCOUNT code can accommodate up to 7 station types, of which 5 can be allocated to any electricity producing system; the remaining 2 are for breeder systems in which the bred fuel is stock controlled. The bred fuel is usually plutonium but in principle the performance of other systems producing fuels such as hydrogen might be simulated.

We have allocated the 7 types as follows:

(1) Conventional fossil fuel systems.
(2) Advanced non-nuclear systems.
(3) Existing nuclear installations.
(4) Advanced thermal nuclear system Type I.
(5) Advanced thermal nuclear system Type II.
(6) Fast Breeder system.
(7) Advanced Fast Breeder system.

This does not simulate the system precisely because:

- Conventional fossil systems include old stations with low thermal efficiencies. But we assume the same efficiency for all stations. However, the error after about 10 years will be small.
- Existing nuclear systems include (in the UK) two types: Magnox & AGR. These may be simulated by a hypothetical reactor type averaging the properties of both.
- All stations are not, as assumed in the simulation, of the same size or, within a given category, of the same performance. Most real stations differ in detail, cost and size and presumably will continue to do so in spite of efforts to standardise.

However, errors introduced by these simplifications are likely to be small compared with the uncertainties in the growth programmes investigated.

The interest in the use of this approach lies in the ability to compare performances of different technical options against different hypothetical programmes in such a way as to demonstrate their efficiencies, resource utilisation and societal impact. The outputs are energy flows and fuel flows from which may be further derived such flows as heat to the environment, plutonium stocks, resource demands, fission product stocks, SO_2 emission etc.

We have examined a number of technological options against three possible UK scenarios. The three scenarios which we will identify as medium, high and low are:

> **Low** assumes that a structural reorganisation of the UK economy is taking place such that the growth rate of electricity demand declines exponentially to near zero by the end of the century.
> **Medium** represents a moderate growth of consumer demand for electricity at 4·5% per annum. (This compares with an historical growth rate of 5·3% per annum from 1960-1972).
> **High** represents a growth of consumer demand for electricity of 5·5% per annum. This is representative of high growth rate forecasts until recently employed by official bodies.

These growth rates do not refer to the rate of installation of stations but to the demand for electricity.

The technological options considered are:
(a) No more nuclear stations to be built.
(b) Installation of nuclear reactors at a rate compatible with industrial capacity and the retention of fossil fired stations at about the same level of fossil fuel consumption as prevailed in 1958-1960. The SGHWR is used for all the new nuclear stations.
(c) As for (b) but using CANDU in place of the SGHWR.
(d) As for (b) with the addition of fast breeder reactors being introduced initially in 1984 but the major FBR programme being introduced at 1990.
(e) As for (d) with CANDU instead of SGHWR.
(f) As for (b) with high energy cost uranium, based on studies of extraction from sea water.
(g) As for (b) but with the introduction of SGHWR delayed until 1990.

The DISCOUNT programme has, as initial conditions, the historical growth of power stations, including the present mix of conventional and nuclear stations (ie 10% of output from nuclear stations in 1975). The merit order of station operation adopted by the generating boards is also included in the basic input data along with station performance limits and energy cost data. The parametric cost data used for the power stations is summarised in Table 2. The derivation of this data will be detailed in our next paper.

The capital and initial fuel inventory costs of all stations (nuclear & fossil) have been spread over a 5 year period with a sixth year of zero cost and zero output prior to its coming on line. Electrical energy and fossil fuel inputs are retained as separate flows throughout. Clearly it has required some arbitrary decisions on whether some of the smaller inputs should be thermal or electrical. Major items such as separative work and the fabrication of capital components have mainly electrical inputs whereas ore mining and milling in remote areas and the production of some materials (pig iron, cement, for example) have fossil fuel inputs but some substitutability is obviously possible. This convention has the added convenience, given present British policy of importing natural uranium and enriching it, that almost all the electrical costs arise in the UK.

Figures 1A, 1B & 1C show the growth of consumer demand for electricity in the Medium, High & Low scenarios. Added to these demands are the electrical inputs required for additional station construction and for the fuel processing. The total output of electricity that has to be provided to meet the demand must allow for the 11% distribution losses and overheads (see an earlier chapter[5]). This additional requirement is also included in the figures along with the increasing nuclear share of total station output. The nuclear

[5] Chapter 2: 'The energy cost of fuels'.

Table 2. Energy cost input data for 1000 MWe stations at 100% load factor

Station Type	Capital cost 10^6 kWht + 10^6 kWhe	Initial inventory* 10^6 kWht + 10^6 kWhe	Refuelling cost pa* 10^6 kWht + 10^6 kWhe
SGHWR	3300 + 570	250 + 800	70 + 300
CANDU	6200 + 900	115 + 40	100 + 30
FBR	4150 + 1050	—	10 + 7
FOSSIL	1900 + 470	—	22000

*based on high grade uranium ores

installation programmes have average annual growth rates of 10% for the medium, 12% for the high and 7% p.a. for the low scenario over the period 1980-2005.

On the scale drawn here the distinction between reactor types and reactor mixes is hardly detectable. These are considered in detail in Figure 2A which shows the electricity consumed in the provision of capital and fuel in meeting the medium demand using different technological options. We will refer to the combination of fossil & SGHWR stations in the medium growth programme as our reference case. For clarity the results of delaying the introduction of SGHWR until 1990 is not shown in this figure. As might be expected the electrical consumption in this case runs parallel to the fossil & SGHWR (Reference) curve shown, diverging from the all-fossil line in 1990. There is a large electrical input where sea water uranium is used, primarily the energy cost of pumping. In Figure 2B these electrical inputs are expressed as a percentage of total station output. Although, in general, the electrical inputs are a small percentage of total output, the differences do represent significant quantities of energy.

The thermal inputs to the medium programme are shown in Figure 3A for the case where future stations are fossil fuelled. The thermal requirements for the provision of additional capital construction and those consumed by the fossil fuel extraction industries are added to this input. Also shown in the same figure are the thermal requirements for the Reference case. In Figure 3B the thermal input scale has been expanded in order to distinguish between technological options. Again, although the difference in inputs is small relative to the total they do represent significant quantities of energy. Both figures include the option where the introduction of nuclear (SGHWR) is delayed from 1982 to 1990, which, after 1994, has the effect of more than doubling the annual thermal requirements compared with the Reference case.

The Reference case requires less thermal inputs than the fossil plus CANDU option (Figure 3B) but requires slightly more electricity (Figure 2B). Based on present conversion efficiencies, the total input (expressed in either thermal or electrical units) in the two options are almost identical, ie they have the same energy costs. This point is emphasised below where we discuss payback periods.

Figures 4A & 4B show the electrical and thermal inputs required for the three scenarios using the fossil plus SGHWR option. Two curves are drawn for low growth; the full lines show the introduction of SGHWR at the installation rate shown in Figure 1C. The dotted lines show a reduced rate of installation starting at the same time (1982) but having the effect of increasing the number of fossil stations *installed* by about 25%. This has the effect of increasing the thermal input requirements by approximately 55% by the year 2000. The reason for this multiplier effect is that not only are more fossil-fired stations required but they have to meet some base load requirements and, therefore, have much higher load factors.

The annual energy flows from the Reference case are brought together in Figure 5. The electrical output delivered to the consumer is as shown in Figure 1A. It does not include distribution losses and provision of capital and fuel. Strictly speaking, the construction and fuel industries, as part of the national economic structure, should be regarded as consumers but we are here concerned with the provision

Energy analysis of a power generating system

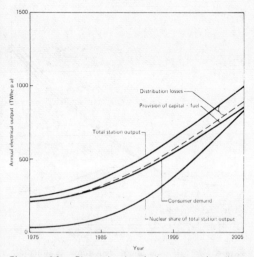

Figure 1A: Planned electrical output (medium scenario)

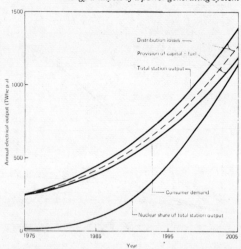

Figure 1B: Planned electrical output (high scenario)

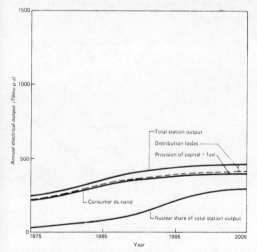

Figure 1C: Planned electrical output (low scenario)

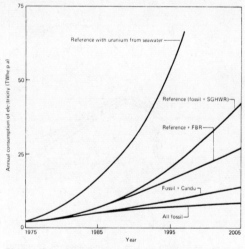

Figure 2A: Electrical input requirements (medium scenario)

of power to other consumers. The two thermal input curves shown in Figure 5 are the all-fossil option and the Reference case taken from Figure 3A.

The general impact of the different scenarios can now be assessed in terms of their effect on fossil fuel usage. The outputs and inputs displayed in Figure 5 are brought together as a ratio which we call fossil productivity and define as:

$$\text{Fossil productivity} = \frac{\text{Electricity delivered to consumers}}{\text{Fossil input}}$$

115

Energy analysis of a power generating system

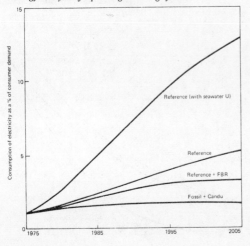

Figure 2B: Electrical input requirements (medium scenario)

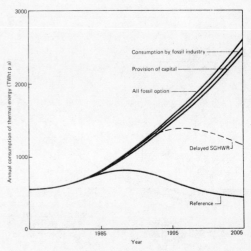

Figure 3A: Thermal input requirements (medium scenario)

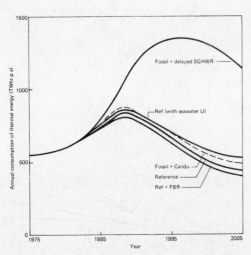

Figure 3B: Thermal input requirements (medium scenario)

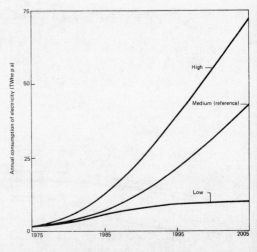

Figure 4A: Electrical requirements for provision of capital and fuel

which has dimensions of electrical energy per unit thermal energy. Figure 6 is a plot of fossil productivity against time for the different scenarios. The dip in productivity which occurs post-1985 in the cases where new nuclear stations have not been constructed is due to the withdrawal of early Magnox stations. Our scenarios assume that all AGRs will be commissioned before 1979 and are therefore, included in the productivity estimates. Any changes to the AGR programme will affect the shape of the dip.

In deriving the energy flows we have not considered the need to

Energy analysis of a power generating system

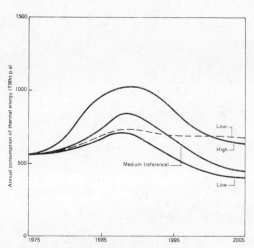

Figure 4B: Annual thermal requirements

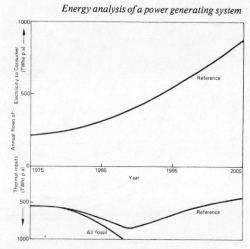

Figure 5: Annual energy flows in medium scenario

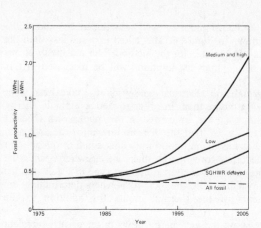

Figure 6: Annual fossil productivity

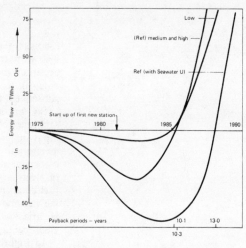

Figure 7: Net energy flows in nuclear component of programmes

build additional enrichment plant to meet the increasing demand although we have included the direct energy inputs to process the fuel. A report in the 'Financial Times' (8 March 1975) suggests £1 000M for the cost of a 10,000 te plant being planned by URENCO-CENTEC for 1985. If the UK share of this is one third, it would be equivalent in financial terms to about one and a half additional SGHWRs with zero output. Spread over the period 1980-1985 this would increase the energy inputs by about 800 GWht and 300 GWhe per annum at a time when the flows as shown in Figure 5 for the Reference case are over 600 TWht in and 250 TWhe out per annum. It

117

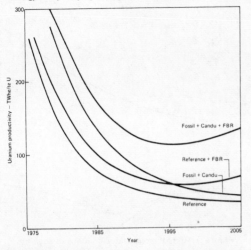

Figure 8: Annual uranium productivity (medium scenario)

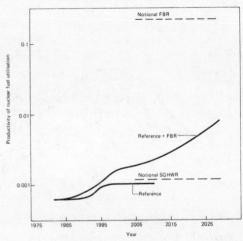

Figure 9: Nuclear fuel productivities (medium scenario)

follows that the fossil productivity will be reduced by about 2% below the levels shown in Figure 6. This effect is much less than the accuracy that could be claimed for the method. The sensitivity of our analysis to our energy input assumptions will be discussed in our subsequent paper.

So far the thermal and electrical energy flows have been kept separate. We now combine them in order to derive a single energy input indicator. Since we are concerned in this paper with efficient generation of electricity we treat the thermal inputs for the provision of capital and fuel as having an opportunity cost equal to the amount of electricity immediately foregone by their use elsewhere. Since the fuel could have been used in the most efficient type of fossil stations we assume a conversion efficiency of 40% when determining the electrical equivalents of the thermal inputs for addition to the electrical inputs.

In Figure 7 we show the net energy flows in terms of equivalent electrical units into the new nuclear investment (ie including Magnox & AGR) of some of the scenarios and technological options, in order to derive the payback period for the nuclear investments. (Note the expanded time base in this figure). Individual reactors go into a positive return on energy invested within a few months of start up, with high grade ores; the precise timing depends very much on the accounting conventions adopted. The payback period for a whole programme, however, exceeds four years after start up, or at least ten years after commencement of construction. Payback would appear to be relatively insensitive to the changes in the modest rates of construction considered here, but is clearly sensitive to the energy cost of the uranium. The high scenario is indistinguishable from the medium in this figure. Eventually the SGHWR electrical inputs exceed the compensating thermal inputs to CANDU and the SGHWR then has a marginally higher energy cost.

In the reference case the maximum total energy deficit in SGHWR construction occurs in 1983 and is equivalent to 12% of consumer demand. In the case of uranium from sea water the maximum deficit arises in 1984 and represents 22% of consumer demand. Electricity generation does not have to be increased pro rata because some of the inputs to the nuclear investment are from fossil fuels.

The form of energy accounting used in this chapter and in the work of others relates useful energy (in this case electrical energy) to essentially thermal inputs. In practice this implies fossil fuel accounting since nuclear systems are not used to provide pure thermal heat as a useful output. The consequence is that although the analysis is not expressed in fossil fuel resource terms the implication is that uranium is a virtually free energy input. Hence the increased productivity which appears to occur as the nuclear component of programmes increases. If the same process is carried out in terms of uranium accounting we have the absurd result that the productivity of uranium falls rapidly as the nuclear component increases, ie there is less 'free' coal, as shown in Figure 8.

The Fast Breeder Reactor (FBR) is proposed not as a means of extending a resource by providing an additional 'free' resource, but as a contribution to increasing the effectiveness of utilising the basic nuclear fuel. Its introduction thus causes the uranium productivity to increase as shown in Figure 8. The FBR has some small effect on fossil fuel accounting as can be seen from Figures 2A and 3B but this is only because it does not require (in the case considered) the extraction and enrichment of fuel.

It is more useful, therefore, to calculate the productivity of utilisation of each fuel separately. In the case of fossil fuels this would reflect improvements in the extraction of fuels, boiler and turbine design and so on, as well as improvements in the energy costs of capital items.

The physics of the fission process decrees that U_{238} is a fuel material since a reaction chain exists to extract nuclear energy from it. Although thermal reactors effectively fission U_{235}, U_{238} fission takes place indirectly from the production and in situ fission of plutonium. Fast reactors on the other hand form an effective way of using the total U_{238} directly. An annual productivity of a nuclear system can, therefore, be defined as:

$$\text{Annual productivity} = \frac{\text{Electrical output} - \text{electrical input}}{\text{Total potential fission energy of HA}}$$

where HA is the number of heavy atoms

Figure 9 shows the annual nuclear fuel productivities calculated. The heavy atoms required to feed the system will depend on the reject level. In the case of enriched reactors, reprocessed uranium is recycled to the separation plant which fixes the reject level. For natural uranium reactors the uranium rejected is whatever comes out of the reactor.

In the case of fast breeder reactors which could be fed with pure U_{238} and by-product plutonium, repeatedly recycled, there is only a small reject of potential fuel. Notional productivities of thermal systems can thus be calculated as follows, (ignoring investment and fuel preparation energy costs and the production and in-situ fission of plutonium):

$$\frac{(\text{Nat.}U_{235}\text{ content} - \text{Reject }U_{235}\text{ content})}{\text{Total Heavy Atoms}} \times TE \times DE$$

Where: TE = Thermal Efficiency

DE = Distribution Efficiency, ie fraction of electricity generated which reaches consumer.

Typical values, assuming distribution losses of 11% are:

Reactor Type	Reject U_{235}(%)	Approximate thermal efficiency	Notional productivity
Gas cooled enriched	0·25	0·4	·0017
Water cooled enriched	0·25	0·3	·0012
Nat. Uranium (CANDU)	0·22	0·29	·0013

An equivalent notional productivity for fast breeder reactors (sodium cooled) is given by:

$$\frac{\text{Total HA Burnt}}{\text{Total HA.Fed}} \times TE \times DE$$
$$= 1 \times 0.4 \times 0.89 = 0.36$$

The approach of the performance of real reactor systems to these values depends on capital and fuel energy costs, their efficiency in forming and in-situ burning of plutonium, and timing of the programme relative to the time they take to reach equilibrium, and, in the case of the Fast Breeder, the doubling time.

It can be seen on Figure 9 that the SGHWR programme shown approaches an asymptote close to the notional value whereas the SGHWR and Fast Breeder, although climbing steadily, has not yet done so at the end of period considered. These performances are not entirely independent of the fossil fuel consumption since this affects the timing and approach to equilibrium.

The attraction of fossil fuel accounting sprang originally from resource considerations. Energy analysis, if it is to have something useful to say, must make a contribution to resource management. We have attempted here to place energy analysis in a dynamic framework but have not tried to provide a basis for resource management. To achieve this requires an objective function of resource utilisation related to the efficiency of providing a service to the life of all relevant resources. This will require new concepts and conventions.

Energy costs of inputs to nuclear power

Earlier in this chapter we showed the energy flows over time for three rates of growth of consumer demand for electricity, with a mix of fossil and nuclear power stations and with different types of nuclear reactors. We adopted the systems approach to our study because we felt that previously published analyses of the energy cost of nuclear power did not present the whole picture. We are not saying that the analyses were incorrect, or invalid, although we may have disagreed with some of the conventions and minutiae. What was lacking was some scale by which to judge the impacts.

Like any large project, the investment in power stations is in deficit for some initial period. How long it is in deficit clearly matters, but 'how much is the deficit' and 'what proportion of available assets does it represent' are other questions that need to be answered. These are the questions to which we addressed ourselves. We have not considered whether the projected consumer demand is necessary, or whether it could be met by technologies not using fossil or nuclear resources.

In this section we present the derivation of the energy cost data that were used earlier in the chapter. These data were summarised in Table 2, p 113.

The accounting conventions we have adopted are as follows:
(1) All thermal and electrical flows are kept separate wherever possible.
(2) The efficiency of the present generating system in converting fossil fuels to electricity is taken to be 25%.
(3) When necessary to combine flows, the conversion of thermal inputs to electrical uses a conversion factor of 0·4. This is on the premise that all future fossil stations will have an efficiency of 40%, so that future fossil inputs to nuclear power construction have an electrical opportunity cost based on this efficiency.

The energy cost of constructing power stations has been derived from a breakdown of the quantities of materials used in their construction. Table 3 shows the quantities of the major materials consumed in constructing a 1000 MWe SGHWR, alongside which are the energy costs of manufacturing these materials. Multiplying these two columns gives the energy content of the materials appropriated in the construction. The cost of fabricating these materials has then to be added in order to arrive at the energy cost of construction. In some cases the split between thermal and electrical inputs is fairly clear, but in others the split had to be made somewhat arbitrarily, after reference to the manufacturing process.

Table 3. Capital construction energy costs (1000 MWe SGHWR).

Material	Quantity	Energy cost per unit		Energy cost of materials	
		(10^3 kWht)	(10^3 kWhe)	(10^6 kWht)	(10^6 kWhe)
Mild steels[a]	40 000 te	9	+ 0·5	360	+ 20
Stainless steel + zircaloy[b]	10 000 te	9	+ 5·5	90	+ 55
Copper[c]	2400 te	10	+ 3·5	24	+ 9
Lead[c]	750 te	13		10	
Miscellaneous metals[d]	4000 te	9	+ 0·5	36	+ 2
Cement[e]	50 000 te	1·9	+ 0·1	95	+ 5
Sand & gravel[h]	200 000 te	0·02		4	
Water[f]	10^7 galls.		$3 \cdot 10^{-6}$		30
Total for construction materials				620	+ 120
Fabrication[g]	60 000 te	14	+ 4	840	+ 240
D_2O	260 te	6400	+ 700	1665	+ 182
Total (including 5% temporary works)				3300	+ 570

[a] NEDO (Reference 14) gives 38·7 GJ, of which 16% (ie, 6·2 GJ) is electricity, to produce 1 tonne steel. Hence energy cost = 9 MWht + 0·5 MWhe, ignoring capital costs, which are a small part of such an energy intensive industry. This compares with Makhijani & Lichtenberg (See Reference 16) who derive costs of 12·6 MWht/te.

[b] NEDO also state that an additional 18 GJ (~ 5 MWhe) of electricity is required per tonne of stainless steel. We have included zircaloy in the stainless steel total; although it has a slightly higher energy cost than stainless steel the quantities involved are only about 4% of this sub-heading.

[c] See P.F. Chapman, *The energy cost of producing copper and aluminium from primary sources*, Open University Energy Research Group, ERG 001, December 1973

[d] Miscellaneous metals, these consist of an assortment some specified such as borated steels and others described by usage such as conduit, switchgear etc. We have arbitrarily assumed that they have the same energy cost as steel.

[e] See References 12 and 14

[f] See M. Slesser and G. Leach, 'Energy equivalents of network inputs,' Dept of Pure and Applied Chemistry, University of Strathclyde, Scotland

[g] See Appendix, p 124

[h] See Reference 16

The energy input to fabrication and to the production of sodium and heavy water are considered in the Appendix (p 124).

The total energy inputs to construction have been increased by 5% to allow for temporary works. It should not be inferred from this that we believe our energy analysis of the inputs to have an accuracy better than 5%, it has been included merely for completeness. The energy costs of constructing CANDU, FBR and fossil stations have been derived in a similar manner.

As can be seen from Table 3, the major energy inputs arise from the heavy water, fabrication, steels and cement. The energy cost of D_2O is almost entirely made up of the direct process energy consumed in the well established H_2S process and is not, therefore, open to much uncertainty. The energy cost of steels and cement are also fairly well established — variations in estimates result from different process efficiencies or different accounting conventions. The material requirements have been obtained from a bill of quantities for a specific reactor design, but alternative designs will probably vary in some degree with, perhaps, a trade off between steel and concrete. The energy input estimate open to most uncertainty is fabrication, which represents 25% of the total thermal input and 40% of the total electrical input to construction of an SGHWR.

If we assume that the overall efficiency of electrical generation is 25%, the gross energy input to the construction of a 1000 MWe SGHWR is 5560 GWht, equivalent to 5560 kWht per installed kWe of capacity. The 1973 capital cost of an SGHWR was £195 per installed kWe[6] which gives an energy cost of capital of 28.2 kWht/1973£ and at 1968 prices (based on the change of purchasing power of the £) is 40 kWht/1968£. Wright[12] derives an energy requirement for capital goods of 44 kWht/£ from his input-output analysis of the UK economy in 1968. We attach no great significance to this coincidence of derived energy costs since the comparison is partly self-fulfilling.

We have assumed that the initial fuel inventory is obtained directly from natural uranium ore deposits, whereas some recycled uranium is included in the refuelling make-up. We have also assumed a common energy requirement for fabrication and reprocessing of all types of fuel elements. The derivation of the energy costs of the fuels is shown in Table 4.

It would be possible but extremely tedious to investigate the sensitivity of our analysis to each assumption and input data. We could also make a detailed comparison of our input data with those of other authors. However, we restrict ourselves here to what appear to be the most significant points.

What is regarded as significant depends to some extent upon a view of the purpose of this study and the usefulness of energy analysis. The conventions we have adopted in our analysis are consistent with our purpose of showing the energy investments required at the present time with existing technologies. If we did the energy analysis in an all-electric economy then the thermal inputs would be obtained from electricity with a 100% efficiency of conversion to heat. The net energy flow will be modified as shown in Figure 10 which is based on the nuclear component of the moderate growth (4.5% pa) of consumer demand considered earlier in the chapter (see Figure 7). The effect of this change of conventions is to increase payback

Table 4. Energy costs of fuel.

1000 MWe SGHWR

	10^6 kWh t	10^6 kWhe
Initial inventory		
620 te U_{nat} (0.32 × 10^6 kWht + 0.016 × 10^6 kWhe per te)[a]	200	10
Enriched to 160 te (2.0 % U_{235})[b]		750
Fab + reproc 160 te (0.25 × 10^6 kWht + 0.17 × 10^6 kWhe per te)[c]	40	30
	250	800
Refuelling per 1000 MWe year		
175 te U_{nat} (0.32 × 10^6 kWht + 0.016 × 10^6 kWhe per te)[a]	60	3
Enriched to 50 te (2.2% U_{235}) (with recycling)[b]		290
Fab + reproc 50 te (0.25 × 10^6 kWht + 0.17 × 10^6 kWhe per te)[c]	12	9
	70	300

1000 MWe CANDU

Initial Inventory		
200 te U_{nat} (0.57 × 10^6 kWht + 0.186 × 10^6 kWhe per te)[d] (including fab + reproc)	115	40
Refuelling per 1000 MWe year		
170 te U_{nat} (0.57 × 10^6 kWht + 0.186 × 10^6 kWhe per te)[d]	100	30

1000 MWe fast reactor

Initial inventory — costs have been included in the fabrication and reprocessing costs of the thermal reactor fuel cycles.

Refuelling per 1000 MWe year		
Fab + reproc 40 te (0.25 × 10^6 kWht + 0.17 × 10^6 kWhe per te)[c]	10	+7

1000 MWe fossil fuel station

Fuelling per 1000 MWe year $\dfrac{10^6 \times 24 \times 365}{0.4 \times 0.99}$	22 000	

At 40% conversion efficiency and 99% efficiency for the fossil fuel industry

[a] The energy cost of 0.3% ores as derived by Chapman (Reference 7). Where uranium from seawater is assumed, we have used values of 4.3 × 10^6 kWht + 2.7 × 10^6 kWhe per te uranium as derived by Taylor & Walford in 'Uranium from Seawater,' PAU.R14/74, Programmes Analysis Unit, December 1974
[b] Assumes 2.42 kWhe/kg separative work. Tails assay of 0.25% U is assumed.
[c] Fabrication and reprocessing costs are assumed to be the same for all reactor systems and are taken from the study by Taylor & Walford (see a above)
[d] Energy costs per unit is the sum of ore plus fabrication and reprocessing costs.

periods from four years to seven years after commissioning of the first station. If very low grade ore is used (ie, from seawater) then the payback period is increased from seven years to nine and a half years and the maximum energy deficit is increased threefold, albeit a very small percentage of total installed capacity in an all-electric economy.[8]

Table 5 shows three 'snapshots' of the nuclear power component in the same building programme of mixed fossil and SGHWR stations (which we earlier identified as the reference programme). The largest term in the energy investment each year is the thermal input to capital construction, 50% of which arises in the production of heavy water. In the case of CANDU, with a larger heavy water inventory, this factor is even more significant. Fortunately, the energy cost of heavy water based on the H_2S process is fairly easily established (see Appendix, p 124) although, of course, it is open to technological change.

The other significant input is the electrical component of the fuel inventory and, when the programme is well established, of refuelling. The dominant factor here is the enrichment process which is also open to technological change by the centrifuge process, although this is not yet sufficiently well developed to have any significant impact during the first ten years of the programme. There is no doubt that power requirements of 2·42 kWhe/kgSWU is the appropriate conversion factor for the enrichment process. What is less certain is the number of separative work units required per reactor inventory and per reactor-year for refuelling. This depends upon the quantity of fuel required, the level of enrichment (which is not constant through the core), the amount of recycling of fuel and also the assay of the waste tailings. Some of these parameters change only with changes in reactor design; in the case of commercial SGHWR these had not been finalised when our analysis was performed, so we took the most likely data at the time. Some parameters change with operating experience and some can be varied to meet changes in fuel strategy and prices. For example, there is a trade-off between natural uranium feed plus recycled fuel against enrichment costs and tails assay (which

Table 5. Annual energy requirement in nuclear component

	1980	1985	1990
Number of stations (625 MWe SGHWR)			
Under construction	16	52	79
Operating	0	8	40
Commissioning	0	4	15
Energy investment (TWht + TWhe)			
Capital	6·6 + 1·1	21·5 + 3·7	32·5 + 5·6
Fuel[a]	0·5 + 1·6	1·6 + 5·2	2·5 + 7·9
Refuelling[a]	0 + 0	0·4 + 1·5	1·8 + 7·5

[a] With high grade uranium

amounts in energy terms to a trade off between thermal and electrical inputs!).

The enrichment of the initial inventory of a commercial SGHWR has been quoted at both 2·0% and 1·8% U_{235}. If we had used the 1·8% figure our initial inventory energy cost would have been 220×10^6 kWht + 650×10^6 kWht instead of $250 \times 10^6 + 800 \times 10^6$. This represents a 20% reduction in energy cost of the inventory which would reduce the calculated energy investment in 1985 (Table 5) by 5%. The effect upon the nuclear system energy payback period is trivial, about 3 months. We therefore feel that our analysis is fairly robust in that the payback period is not likely to be modified significantly by realistic modifications to our data. The two reactor types we have examined were chosen because we had access to the input data and because they enabled us to compare a natural uranium system with an enriched system.

We have not made detailed com-

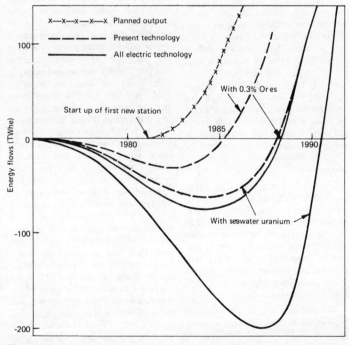

Figure 10. Net energy flows in the nuclear component of a power generation programme

parisons with other reactor types in the analysis partly because we could not be certain that the input data would be as reliable as in the cases examined. For the benefit of those readers who believe that useful comparisons can be made, Chapman's data[7] for PWR capital + fuel inventory costs cover a range from 10% higher than our SGHWR data to 20% lower. Applying these data to the building programme in 1985 (Table 5), in the same way as discussed above for changes in fuel costs, could modify the payback period by as much as +6 months or −9 months. This simple calculation is merely illustrative. Because the fuel performance of a PWR is different from an SGHWR, the rate of installation of stations will not be the same, although it is unlikely to be very different over the payback period.

The casual observer of the protracted debate on the energy analysis of nuclear power can be forgiven for being confused. He may have heard, or read, that the payback period for nuclear power is 2-3 months from start up, or 8-24 months, or that payback occurs after 4 years, 9 years, 15 years and even never. We can add to the confusion by demonstrating that all the above answers can be derived from an agreed single set of input data merely by adopting different definitions for nuclear power (ie, by referring to a single reactor or to an extended building programme), by adopting different conventions (ie, by excluding or including transmission losses, by the use of an opportunity cost for fossil inputs, or by assuming a one-to-one equivalence for thermal and electrical energy), and by adopting different rates of growth of planned output.

The use of energy ratios or power ratios as performance indicators can be made equally confusing by the adoption of different conventions. Any attempt to rank reactor systems by means of some single figure of merit is bound to be misleading; rankings may alter with different assumptions. Energy analysis is not sufficiently precise to be used reliably for ranking purposes.

The problems are no different from those of the investment analyst who has an even wider choice of indicators: payback period, return on capital, internal rate of return, net present value, and so on. All of these can be used with different accounting conventions relating to assumed plant life, discount rate etc. It is only by drawing together all these factors and examining them in relation to the decision criteria that they become useful. In the UK public sector the Treasury give guidance on the accounting conventions, but the decision criteria are more susceptible to political objectives. There are no such guidelines for energy analysis. Some conventions are based on technology and thermodynamics but the laws of thermodynamics are not consistent with consumer behaviour — some units of energy are more valuable than others and the valuation probably changes with time.

By using it within a systems framework, some of the weaknesses of energy analysis are shown to be less important than one might have expected. A feature of its use in this present study shows that the spectacular conclusions relating to payback periods which have been arrived at by partial energy analysis are not in evidence in a systems study.

We do not see that energy analysis has any policy making role. Fortunately many other energy analysts have also recognised that analysis of a single resource (albeit in several forms) is not a strong decision making tool. We agree with Chapman[9] that energy analysis demonstrates that nuclear technology is a better way of producing electricity from fossil fuel, but this strong statement becomes much weaker when we add that present nuclear technology is also an inefficient user of nuclear fuel. To us, energy analysis is only one example of resource analysis that should be used to show the impact on scarce resources of alternative decision options. In our analysis we show the energy investments in relation to present UK levels of consumption but not in relation to total reserves. Our analysis does not tell you whether there are other constraints that prevent these levels from being achieved. It is to these areas that the analyst should now be looking. There is satisfaction to be gained from resolving some of the outstanding problems of what energy accounting conventions should be adopted, but there is no merit in improving the third significant figure when the second one is in doubt.

Appendix

We include here our derivation of the energy requirements of heavy water and reactor grade sodium, as well as for fabrication of capital items.

(i) *Heavy water* — based on the H_2S process which requires 5600 kg steam at 60 atmos per kg D_2O output. There is an electrical requirement to drive pumps and compressors. The direct energy inputs = 6.0×10^6 kWht + 0.7×10^6 kWhe per te/D_2O.[10,11] The estimated capital cost = £120/kg D_2O p a in 1972.[10]

Assuming a ten year plant life and an energy cost of capital of 33 kWht/1972£[12], then the capital energy requirement = 0.4×10^6 kWht/te D_2O.

The energy costs of the H_2S and process water are insignificant in comparison with the above figures. Hence we conclude energy requirements of heavy water = 6.4×10^6 kWht + 0.7×10^6 kWhe per tonne

The D_2O make up for SGHWR is estimated to be less than 0.5% of its inventory per annum, which represents about 5% of the energy cost of refuelling and has not been considered in our analysis.

(ii) *Reactor grade sodium* — based on process data. The data shown in Table 6 include capital and running costs, including purification to reactor grade. The dominant term is the electrical input for electrolysis. We have given an energy credit for the chlorine produced which makes a significant contribution to the energy requirements.

Table 6. Energy requirements for sodium (Reactor grade)

	Per tonne Na	
	(kWht)	(kWhe)
Capital	200	50
Running Costs		
Sodium chloride	4500	50
Sodium hydroxide		50
Electricity		10 000
Steam	300	
Natural gas	1000	
Water	4	
Nitrogen		200
Miscellaneous salts	300	100
Chlorine (credit)	(11 300)	
	(5000)	+ 10 400
Transport	600	
Reprocessing	1600	

(iii) *Fabrication* The 1971 input-output tables[13] show, for the construction industry (SIC 500), that total material inputs (excluding income, profits, taxes) were £2859 million of which £60 million is from the energy sectors (£17·6 million from electricity). Assuming an average price of 0·1 p/kWht in 1971, direct energy input to the construction industry = 60 x 10^9 kWht. The final output of this sector was £5718 million. Hence consumption was 12 kWht direct energy per £ value of output.

The capital cost of a 1000 MWe SGHWR in 1971 was about £150 million. At 12 kWht/£ this implies a direct energy input of 1800 x 10^6 kWht to fabricate 60 000 te's of material, ie 3 x 10^4 kWht/te. Split between thermal and electrical inputs, the energy cost of fabrication = 14 x 10^3 kWht + 4 x 10^3 kWhe/te material.

NEDO[14] state an estimated average cost for processing nonferrous metals of 11 x 10^3 kWht/te and for ferrous a range of 1·5-6 x 10^3 MWht/te. Berry & Fels,[15] in their automobile study suggest 2-9 x 10^3 kWht/te for processing of basic materials (rolling, forging, extruding etc.) plus 6 x 10^3 kWht/te for fabrication, giving a total of 6-15 x 10^3 kWht/te for fabrication. This is less than the above figure for the construction industry. It is also less than the industrial plant & steelwork industry sector (SIC 341), which, from the input-output tables, consumes £11 million on direct energy inputs (of which £5·2 million is on electricity), and has a total output of £1050 million, leading to a direct energy consumption of 10 kWht/£, which if applied to the SGHWR cost of £150 million gives a figure of 25 x 10^3 kWht/te.

Since nuclear power construction makes considerable use of both the construction industry and the plant and steelwork industry we have chosen to adopt the higher figures rather than the data for more specific materials.

[6] Department of Energy, 'Report on the Choice of Thermal Reactor Systems,' Nuclear Power Advisory Board (1973)
[7] Chapter 9: Energy analysis of nuclear power stations'.
[8] This extreme case is clearly unrealistic since some of the thermal inputs are chemical feedstock for which electricity is no substitute.
[9] P.F. Chapman, 'Conventions, methods & implications of energy analysis,' *Understanding Energy Systems*, Inst Fuels & OR Soc, April 1975
[10] S. Villani, *Energia Nucleare*, Vol 21, p 139 (1974)
[11] L.R. Haywood *Nuclear Engineering International*, June 1974
[12] D.J. Wright 'The Natural Resource Requirements of Commodities' *Applied Economics* Vol 7, p 31 (1975)
[13] *Input-Output Tables for the UK 1971*, HMSO, 1975
[14] *Energy Conservation in the UK*, NEDO, 1975
[15] R.S. Berry & M.F. Fels 'Energy Cost of Automobiles' *Science & Public Affairs*, December 1973
[16] A.B. Makhijani & A.J. Lichtenberg 'Energy & Wellbeing,' *Environment*, Vol 14 No 5 (June 1972)

12. The economics of energy analysis

Michael Webb and David Pearce

Energy analysis has assumed an apparent major importance in the planning of energy use. The authors argue that debates over the detailed aspects of energy analysis and its applications are misplaced in face of the absence of any real assessment of the purpose of energy analysis. They offer the view that energy analysis fails to achieve its stated purposes and that economic analysis already provides a rational base for planning energy use, making energy analysis either superfluous or misleading.

This chapter is a shortened version of PSERC Discussion Paper 75-03 available from the University of Leicester

Authors' note
We are heavily indebted to Peter Chapman of the Open University for his generous help in supplying documentation and for an exchange of views in private correspondence. If Dr Chapman's name figures prominently in this chapter it is not a reflection of any victimisation on our part but of his prolific input to the energy analysis literature!

[1] National Economic Development Office, *Energy conservation in the United Kingdom: Achievements, Aims and Options*, HMSO, London 1974. See especially pp 90-1.

The object of this chapter is a critical appraisal, from the economist's standpoint, of what appears now to be called energy analysis, but which has, at one time or another, been called energy budgeting, energy accounting and energy costing. The proliferation of articles on energy analysis and the fact that it has apparently been afforded serious attention in political circles and official documents,[1] are indications of the importance now attached to this technique. Remarkably, however, energy analysis has been subjected to only a minute amount of published criticism, although we are aware of extensive verbal criticism from many quarters. We seek to correct the balance as far as the published literature is concerned. While what we have to say is frequently critical in a purely negative sense, we hope that what we have to say will prompt energy analysts to define in a rigorous fashion the real uses of their studies.

The genesis of energy analysis

It is useful to begin by asking why energy analysis should, in recent years, have become so important, even if that importance, in our view, is exaggerated. The genesis of EA is clearly the awareness, now a commonplace, that the world has many natural resources in finite supply. Of these, energy resources — at least as far as the fossil fuels and uranium are concerned — are clearly limited as a stock. Others such as solar energy are limitless flows (within any sane time horizon anyway). However, as far as the 'renewable' or unlimited fuels are concerned, technology is not developed to apply them to practical use and they must remain in the realm of probabilistic future supplies. To plan the future on the basis of what *might* be is to engage in a maximax policy the costs of which, if technology fails to generate the required new fuels, would be catastrophic to future generations.

We accept that the critical issue in what we might call non-speculative planning is that of rationing resources intertemporally in the most equitable fashion. Inter-generational equity is a subject of detailed concern by economists, although we would be dishonest if we suggested that all economists believe in accommodating the problems bequeathed by one generation to another in their analysis. But what we *can* say is that energy analysis offers no assistance with the problems of intertemporally allocating resources in finite supply.

The reason for this is simple — EA is a purely mechanistic technique devoid of all value content (even though, as we shall argue, its exponents often use it for evaluative purposes). As such it can tell us nothing about optimal allocation rules. We make no specific counter-claim for economic analysis either. The literature on the economics of intertemporal allocation is frequently rarified and this is not the place to discuss its merits.[2] What we are concerned to note is that the *fact*, if it be one, that current technology sets a limit to the availability of exploitable energy is not one that affords EA any importance. Instead, it seems to us to make the normative issue of intertemporal allocation that much more important, and this is not something that EA can, in our view, assist.

Slesser[3] writes that 'economics treats the world as a closed system having access to limitless amounts of energy, whose acquisition takes only time, capital, labour and technology'. But this is simply false. There is nothing in economic science that requires us to assume limitless resources of any kind. Scarcity is, in fact, the very foundation of economics.

The same false charge that economics assumes 'no shortages of any inputs to the production system' has been made by Chapman.[4] He compounds the error by stating that economics assumes substitutability between inputs whereas energy analysis does not. Indeed, both Slesser and Chapman emphasise that, if capital and material inputs are reduced to energy units, the result is a two-input model of the economic system — the two inputs being labour and energy — then there will be non-substitutability between labour and energy. As Slesser says 'in the last analysis, energy does what labour cannot do'. Continuous substitutability between inputs is indeed an assumption of neoclassical economic analysis, but it is not a necessary one and much of the progress in economics since the formulation of the neoclassical axioms from 1870 to 1940 has been in the realm of reassessing the neoclassical results in the context of product discontinuity. The general outcome of this analysis is that the original results of the neoclassical system remain intact.

However, even if the possibilities of substitution are non-existent, it remains the case that the problem again reduces to one of rationing resources over time. Thus, if labour and energy are, in some sense, the only resources, if substitution is constrained, and if energy is a finite non-renewable stock whereas labour is a renewable flow resource, the policy options are:

- Allocate resources over time according to some intergenerational welfare criterion.
- And/or restructure the configuration of material output in favour of labour-intensive and against energy-intensive outputs.

It is evident, and would seem to be admitted by most energy analysts, that EA has nothing to tell us as far as the first option is concerned. Its role in the second also seems more than questionable. If energy is scarce then *some* price change will take place, generating exactly the product substitution called for. However, if, as we would accept, market prices are non-optimal even with respect to current-generation biased decision-rules (and hence even more non-optimal with respect to future-oriented rules) then it would be useful to identify energy intensive activities so as to adjust market prices to reflect the true shadow price, perhaps by the use of an energy tax.

[2] For an interesting contrast see the essays by I.F. Pearce and by J. Kay and J. Mirrlees in D.W. Pearce (ed), *The Economics of Natural Resource Depletion*, (Macmillan, Basingstoke, 1975).
[3] M. Slesser, 'Accounting for energy', *Nature*, vol 254, March 20, 1975.
[4] Chapter 1: 'Energy costs: a review of methods'.

What is not clear, however, is why we need EA to identify such energy-intensive uses: a tax on energy consumption can be implemented without carrying out elaborate exercises to identify energy use. If, say, some tax proportionate to energy consumption was introduced, energy-intensive activities would *automatically* bear the heaviest tax burden, simply because energy costs comprise part of the costs of production of economic activity and because these costs are shifted forward from the most basic economic sectors such as resource extraction to the final product. In short, we fail to see where the 'two factor' approach adopted by energy analysts takes us. At best it adds nothing to what simple economic analysis tells us, at worst it serves only to obscure the issue.

We may also note at this point that reducing inputs to energy and labour obscures the fundamental reason for the separating out of capital by economists. Basically, capital generates a flow of goods in excess of the original value of the capital. Indeed, this is the very rationale of capital investment. If it were not, we would merely be diverting resources from one use to another with the same value and hence gaining nothing. We have nowhere seen in the EA literature reference to the productivity of capital.

Clearly it can be accommodated in the sense that capital reduced to energy units generates a flow of energy values if the project is productive of energy (eg a nuclear power station). We would then have a situation in which there would be labour inputs (measured in man-hours or in value terms?) plus energy inputs to be offset against energy outputs plus any non-energy outputs. But such a calculus is devoid of any use, because it offers us no decision rule by which to choose options. If we have a project using a combination of, say, 4 energy units and 6 labour units, with output 5 energy units and 3 units of some commodity, how do we decide if it is worth undertaking?

The absence of a homogeneous measuring rod relating to values in energy analysis makes such a calculation worthless. The only way homogeneity of units could be achieved would be to reduce labour to energy terms as well and to value commodity output in energy terms. The latter appears to be what some energy analysts would want to do, but they resist the reduction of labour to energy units. As such we have no homogeneous unit and hence no decision criterion.

Alternatively, the rule might be to select projects which minimise energy costs regardless of the levels of other inputs. Energy analysts appear divided on whether such a rule is sensible. While protesting on the one hand that EA cannot be *evaluative* it is not in the least difficult to find in the literature statements such as 'energy analysts believe that it makes sense to measure the cost of things done, not in money, which is after all nothing more than a highly sophisticated value judgement, but in terms of thermodynamic potential.'[3] But if EA is not evaluative, what is the point of such an elaborate exercise?

We elaborate on this point in later sections. For the moment we argue that the energy analysts' attempt to differentiate their subject from others by reducing economic systems to two-input analysis serves no useful function and, indeed, only obscures some important aspects that differentiate energy embodied in capital from energy embodied in other commodities.

There may well be other 'energy limits' or 'boundaries' to economic activity. Many writers have commented on the pollution limits set by the effects of waste heat dissipation. Georgescu-Roegen for example states 'The additional heat into which all energy of terrestrial origin is ultimately transformed when used by man is apt to upset the delicate thermodynamic balance of the globe.'[5] These expressions of concern are well-taken and do indeed reflect lack of attention to the laws of thermodynamics by economists. What is not clear, however, is why we need EA to identify this boundary, or, if we do need it, why the existence of such a boundary is thought to involve some deep criticism of economic analysis.

The most basic assumption of economics is that economic agents – consumers and producers – seek to optimise subject to constraints. If there is indeed a dissipated heat limit to economic activity this constraint can be added to the optimisation problem. Occam's razor demands that we do not multiply our techniques beyond the minimum necessary if existing techniques are quite capable of accommodating the problem.

The homogeneity assumption and relative prices

Energy analysts treat energy as an entity that can be aggregated *regardless of its source*. The exception is labour energy which is differentiated from other energy inputs.

In energy analysis, any commodity, i, can be 'reduced' to some energy content E_i plus some labour content L_i. The volume of output can be left in physical, monetary or energy terms, giving us expressions of the 'cost' of a unit of that output of the form

$$\text{Energy Cost} = \frac{Q_i}{E_i + L_i}$$

where Q_i is the output measure of i. We have already commented on the difficulties of finding a use for such a ratio. We may note that the aggregation problem in economics is 'solved' by using prices. That is, Q_i would be measured in value terms, and the denominator would appear as

$$\sum_{e=1}^{e=n} P_e . X_e + \sum_{L=1}^{L=m} P_L . L$$

where P_e, X_e refers to the price and quantity of different energy sources, and correspondingly for labour. The prices in these expressions will, if markets function properly, reflect consumers' willingness to pay for the product in question, or, for inputs, the benefit foregone to consumers by using the input in its current use rather than the next best use (the input's 'opportunity cost'). Where markets do *not* operate properly[6] the prices used are *shadow prices* – prices which, if they did operate, would reflect marginal willingness to pay on the part of consumers. Either way, the use of prices serves to homogenise the heterogeneous units *and* to import value-content to the resulting aggregate.

Before proceeding to discuss the validity of the analogous procedure in energy analysis – the use of energy units to homogenise inputs – we may note that shadow prices also bear *some* relation to the finitude of resources. Essentially, if markets operated freely and perfectly, the current price of a resource would reflect expectations about the limited stock of that resource. As the stock is depleted the

[5] N. Georgescu-Roegen, 'Energy and economic myths', *Southern Economic Journal*, vol 41, No 3 January 1975.

[6] 'Properly' in this context relates to a situation in which the configuration of prices maximises consumers' welfare in the aggregate. The EA literature contains numerous comments on the biases imparted by using market prices but we have nowhere noticed even an awareness of the idea of shadow pricing, perhaps because energy analysts mistakenly identify economics with the free enterprise ethic.

price will rise, thus rationing the use of the resource, inducing the adoption of substitutes, encouraging recycling, and so on. Where the *totality* of resources of a specific kind, such as energy, are concerned, this mechanism will not operate except in the sense that labour and capital can be substituted for energy.

Now, energy analysts are quite right to point out that the possibilities of this kind of substitution are limited, although not, one suspects, as limited as is often suggested. We know, however, that markets in natural resources such as energy do not function perfectly. We need not dwell on the reasons for this here[7] but it is important to note that we do not know the extent of the deviation of the appropriate shadow prices from the actual market prices. We have noted a tendency in some of the EA literature to assert that market prices fail *totally* to reflect future scarcity, an assertion that is nowhere substantiated by any evidence. Indeed, assessing the evidence is a complex issue.

We are prepared to believe, however, that current resource prices are not accurately related to future scarcity. What we need to know, then, is how EA will assist us in identifying the future limits. As we argued earlier it is not in the least clear how it assists.

We can now turn to the central matter of this section: the homogeneity assumption in energy analysis. To demonstrate the problems of using a common energy unit (or, indeed, *any* common physical unit) we consider an example which contrasts such a physical measure with the economist's concept of opportunity cost outlined above. Assume the existence of some resource, call it 'oil', which is homogeneous and in finite supply. In addition assume that all the deposits of this resource are equally accessible and thus that there is no change in the physical inputs required to obtain a ton of this resource. Thus each ton can be obtained at a constant expenditure of energy. Finally, for simplicity, assume that there are no substitutes available for this resource.

As the physical exhaustion of this resource approaches, energy analysis will continue to measure its cost calculations at a constant 'cost' (in therms or kWht etc). But economic analysis would show the price of 'oil' increasing to reflect its increasing scarcity. This would probably happen in two ways. Suppliers, seeing the coming exhaustion of their product and knowing of the absence of the possibility for substitution, will raise its price. Second, if there exists a futures market, dealers in this market would offer higher prices for the product as the time of its physical exhaustion approached. If there is no futures market, the price will nonetheless rise because of supplier reaction. Whether the time-profile of prices that results is an 'optimal' profile from the point of view of social welfare is not relevant to this example. We noted above that the profile may well deviate from such an optimal path. *Our point is that prices will rise by some amount to reflect scarcity, whereas the energy cost will be constant.*

In such a situation energy analysis would be a poor guide to increasing scarcity: indeed, in this case it would indicate no scarcity at all. Now, energy analysts have emphasised that one of the main functions of EA is to identify changes in relative prices over time. It is argued that, because of the deficiencies of the market mechanism, economic analysis will not identify those changes, or, if it does, will do so later than energy analysis.[8] Indeed, this appears to be the sense of some of the more grandiose claims for EA. Berry, for example, has

[7] See the introduction to D.W. Pearce (ed), *The economics of natural resource depletion* (Macmillan, 1975), where the various divergencies are listed.
[8] P. Chapman, 'Energy analysis: A review of methods and applications', *Omega* forthcoming.

said 'if economists in the market place were to determine their shortages by looking further and further into the future, these estimates would come closer and closer to the estimates made by their colleagues, the thermodynamicists.'[9] Our simple example shows that no such convergent process need occur.

Now suppose that instead of the 'oil' being equally accessible it has a decreasing quality gradient – to obtain the marginal barrel of oil we need to expend extra amounts of energy and other resources. This is perhaps more pertinent to many material resources. As more and more marginal resources are exploited we can expect the energy cost to rise and hence there is some relationship between scarcity and energy cost. Equally, however, we will find that the real economic cost of extracting the marginal resource will rise. We can illustrate this by taking the example of copper extraction.

We know that the grade of copper ore has been declining. It has been calculated that fuel input per unit of copper output for the USA fell to about 1930, rose slightly from 1930 to 1950 and then rose very fast indeed to 1960. The curve is in fact a flat-bottomed 'U' shaped curve. The important question relates to the information provided by this curve. Would it indicate that energy analysis has pinpointed a rise in the relative (real) price of copper earlier than economic analysis would, and hence is more sensitive to scarcity? It is certainly the case that if we look at the real price of copper (the money price deflated by an index of manufacturing wage rates) we find it rises much later than 1930 – somewhere in the mid 1960's.[10] Has energy analysis therefore anticipated the relative price rise?

The problem is that it is impossible to draw any conclusion at all from such an analysis. Firstly, the energy cost of copper extraction is only one of the costs involved. The energy approach and the economic approach are therefore non-comparable. Secondly, if we took the money costs of fuel inputs we would secure the same result as EA. Quite simply, whatever the energy inputs into copper extraction, and however far back the energy costs are traced through the economic system, those inputs will have prices and the accumulation of prices will be revealed in the final money cost of fuel for copper extraction. In short, *if* we are interested in energy inputs alone it is completely unclear why we require energy analysis rather than the straightforward and readily obtainable money cost of energy inputs.

Thirdly, the analysis *presumes* scarcity rather than demonstrates it. That is, the fact that the energy costs of securing extra copper are rising need in no way be correlated with a scarcity situation. We argued above that a decreasing quality gradient situation would tend to have rising energy costs associated with it. It does not follow from this, however, that every situation in which we observe rising energy costs is a situation of scarcity. In contrast, whatever the defects of market prices, if we observe the real price of copper rising we *can* deduce something about scarcity.

We must be careful not to claim too much for economic analysis in this respect, however, for it is equally true that if markets fail to operate at all sensitively, constant or falling real costs might exist even though a scarcity situation might exist. All we are saying here is that we fail to see how energy analysis improves our knowledge of the situation. Fourthly, simply because energy inputs per unit of output rises *before* the real price of copper rises, no conclusion to the effect

[9] R.S. Berry: US Congressional Record, 92nd Congress, S 2430, 1972. Quoted in M. Slesser, 'Energy analysis in technology assessment', *Technology Assessment*, vol 2, No 3, 1974.
[10] W. Nordhaus, 'Resources as a constraint on growth', *American Economic Review*, 1974.

that energy analysis has *anticipated* a relative price rise can be deduced. The analysis tells us nothing, for example, about technological change. It also assumes a decreasing quality gradient which, while it may be a sensible assumption for copper, is questionable for other resources.

Finally, whereas the economic argument that current prices reflect future scarcity has *some* rationale to it, a rationale based on the maximising behaviour of resource owners, energy analysis offers a purely mechanistic interpretation of future scarcity based on the simple proposition that the exploitation of copper has led to the processing of leaner and leaner ores. To put it another way, what does energy analysis tell us that is not already contained in the fact, readily ascertainable, that copper concentration in ores has been declining over the years? If we reformulate the EA proposition as saying the declining ore quality is a sign of increasing scarcity we are stating the obvious as long as we ignore technological change and the chance of higher quality ore discoveries.

We can extend the analysis further. So far we have considered the case of an homogeneous resource with no decreasing quality gradient, and the case where the quality gradient does decline. Now we consider the case in which we have several different types of energy inputs, say electricity from nuclear power, hydro-electricity and coal. In economic analysis different inputs are held to be efficiently allocated if the extra output (marginal product) produced by each last unit of input used to produce each output is the same. The marginal product of each input must be the same in all its uses. In the economist's language, the marginal rate of transformation is common between all inputs.

Now, different types of energy may be used in various ways — they can be used to supply energy directly, or they can be converted into other forms of energy for indirect use (coal into electricity etc). As Turvey and Nobay have pointed out,[11] the marginal rate of transformation of factors in production (assuming the satisfaction of the conditions required to give an efficient allocation of resources) gives one economic measure of how one fuel type should be converted into each other. In cost terms the various fuel inputs should be valued at their marginal production costs. Where fuels are purchased by consumers the conversion factor is given by the rate at which the consumer is prepared at the margin to substitute one fuel for another so as to maximise his utility. This is the marginal rate of substitution and, assuming efficiency in the allocation of resources, is measured by the marginal cost of the fuel to the consumer.

What these concepts tell us is that in both production and consumption a therm is not necessarily a therm. Fuels have a number of attributes and heat content is but one. Two fuels with the same heat content but other different attributes (in terms of cleanliness, transportability, etc) would have different marginal costs. In terms of economics the fact that two forms of fuel have the same heat content does not make those fuels identical. But in energy analysis the assumption of homogeneity (kilowatt hours are kilowatt hours regardless of how they are produced) obscures this important difference. *The calculation of energy costs using the homogeneity assumption makes energy analysis irrelevant to the process of resource allocation now and over time. If, on the other hand, the*

---

[11] R. Turvey and A.R. Nobay, 'On measuring energy consumption', *Economic Journal*, December 1965.

homogeneity assumption is relaxed, energy analysis has no foundation.

This point needs considerable emphasis since it lies at the heart of the misuse of energy analysis, and is the foundation of its exaggerated importance. Only by assuming homogeneity can EA proceed. But once homogeneity is assumed EA loses all relevance to resource allocation decisions. We may note that this precludes EA from being used for virtually all of the purposes claimed by energy analysts. Examples are numerous. Thus it has been claimed[12][13] that if the average thermal efficiency of power stations is 25% then 75% of the energy input is lost. But within the economic system consumers are demonstrating a preference for a secondary fuel input over a primary fuel input. The price which they pay for electricity will reflect the opportunity costs of the inputs, including coal, used to make it. This is true irrespective of how much of the heat content of coal would be released in its alternative uses.

In the market economy system if consumers demand coal as an input to make electricity they are saying that they value its use in this way more highly than in its alternative uses, even though in these alternative uses all the coal's heat content may be released. It follows that it is misleading to talk of the therms not directly converted into electricity as being 'lost'. Their alternative uses are assessed by consumers when the price system functions reasonably well.

Another problem with the implicit homogeneity assumption can be illustrated using an example given by Chapman.[4] He sees one of the possible uses of energy analysis as being the ranking of alternative energy conservation investment projects. Such projects are to be ranked simply in terms of the number of therms saved per £ invested. Now clearly this implies that all therms are equally 'worth' saving, whether they come from domestically produced coal, imported oil, or foreign enriched uranium.

To appreciate some of the problems involved with this approach consider the following example. Suppose that expenditure of £1 on the enforcement of speed limits led to a saving of 10 therms in reduced petroleum consumption; that an expenditure of £1 on a law limiting the heating of public buildings led to a reduction of coal consumption (used for electricity) of 11 therms and that an expenditure of £1 on the development of a new gas cooker (of better efficiency) led to a saving of 12 therms.

On the simple objective being used the last policy would be the best. But this choice would imply that not only would the alternative use values of the alternative fuels be neglected, but, in addition, it would be assuming that saving a therm of imported energy was equivalent to saving a therm of domestically produced energy. On the first of these points it is quite clear that if a government really wished to do this it could save a considerable amount of energy at little financial cost merely by introducing physical controls on the use of energy, eg rationing. But if such a policy were to consider only its costs of implementation and the resulting energy savings in physical units, it would be ignoring the value of the benefits foregone due to the reduced consumption of energy. The problem then is simply that the saving of energy measured in physical units implicitly assumes a one-to-one correspondence of benefits foregone to energy (per therm) saved. In the market type economy there is no reason why this correspondence should exist.

[12] P. Chapman and N.D. Mortimer, *Energy Inputs and Outputs of Nuclear Power Stations*, Open University Energy Research Group, Report ERG 005, 1974.
[13] P. Chapman, 'The ins and outs of nuclear power', *New Scientist*, 19 December 1974.

Energy analysis as a normative technique: energy conservation

So far we have tried to concentrate on what we might call the 'positive' claims of EA, claims which we feel have not been substantiated. We now turn to the wider claims for EA. These state that EA has some evaluative purpose. We are very much aware that energy analysts, in the main, have declared, in some cases repeatedly, that EA is not evaluative. Thus, Chapman declares 'Energy analysis does not tell anyone what they ought to do'.[13] Since energy analysis is merely an analytical technique this is what we would expect. Unfortunately, however, these same authors have then used this technique to make policy recommendations. The same author has stated that energy analysis can be used to *rank* alternative energy conservation policies.

Used by itself this is just what it cannot do. The ranking of alternative policies must be in terms of some specific objective function, and this function takes us away from the positive aspects of energy analysis into normative issues since this objective is necessarily not part of the analytical method. We might add that Chapman's denial of the evaluative role of EA is not supported by some of his colleagues. Hannon is quite explicit: 'In the long run we must adopt energy as a *standard of value* and perhaps even afford it legal rights'[14] (our italics).

Fundamental to the choice between, or ranking of, alternative policies is the specification in an operational form of an objective function. In energy analysis this is also necessary in order that the boundary of the system should be delineated. In none of the papers on energy analysis that we have read is the question of the form and specification of the objective function discussed adequately. This may be the result of the (correct) view that energy analysis has no normative significance. But, as has been mentioned, in many of these papers policy questions are discussed (and sometimes recommendations made) and so the relevant objective should have been stated clearly. In this connection it is important to note that in economics the costs and benefits of particular actions cannot be defined or measured until the associated objective function has been specified operationally. Market price data are relevant in the pursuit of some objectives, while for others shadow prices must be used. Since the use of money values is sometimes recommended in energy analysis for the choice between alternative policies it follows that the associated objective must be specified.

From the writings of a number of energy analysts it would appear that one of their prime concerns is with the question of energy conservation. They are particularly concerned to ensure that in the development of some energy source (eg shale oil or nuclear power) more energy is not invested that will be produced. It is therefore pertinent to enquire whether energy conservation can be considered to be an (or the) objective.

Expressed in this way the answer must be 'no' because it is non-operational. How much energy is to be 'conserved' over what time period? In what geographical area? Since all methods of production involve the use of energy should the economic growth rate be chosen to maximise the rate of energy conservation? Since even a zero growth rate involves positive production levels the required growth

[14] B. Hannon, 'Energy conservation and the consumer', *Science*, vol 189, No 4197, July 11, 1975.

rate would be negative. We presume this is not what is meant. Perhaps what is meant is that each productive process (when substitutes are available) should be selected so as to minimise the energy requirement. If this is the intention, then general, rather than partial, equilibrium analysis is required and we note with interest that the development of satisfactory physical input-output tables would meet this objective.

It remains the case, however, that objectives such as minimising the energy input of a given output are distinctly evaluative. It introduces the idea that energy as a constraint on economic activity is more important than any other constraint. If, for example, we selected policies on the basis of energy content, preferring those with low energy input to those with higher energy input, we could easily find ourselves in a situation in which we would be adopting policies with high total resource cost.

Certainly, energy conservation can be furthered by switching from energy-intensive products to non-energy-intensive products. The difficulty is to see why we need EA to further this end. Quite simply, the energy inputs into any economic process will have a price attached to them. This price will reflect the resource costs of supplying that input and included in these resource costs will be the energy inputs. In this way, the price of an energy input is built up from all the related previous processes. An energy conservation programme would require knowledge of how much energy costs will be saved by switching between products, information obtainable from a monetary input-output table just as readily as from a physical input-output table of the energy analysis kind. Further, the use of monetary measures would at least offer some indication of social preferences for the commodity switches whereas EA offers us no such guidance, as we have repeatedly pointed out. In short, energy conservation measures based on some index of energy input to commodity output implies an objective function, but it is a function unrelated to consumer preferences and we see no justification for adopting EA as the appropriate technique when a preferable one exists.

Chapman has stated explicitly 'Thus if you want to adopt an effective energy conservation policy you can compare the costs of various policies (in £'s) with the amount of energy saved overall (in kWht's or therms or joules etc). This allows you to choose a best 'buy' in energy conservation'.[15] Chapman claims that this choice cannot be made using economics because markets are imperfect and market-supplied data will be a poor guide to resource costs. In our view there are so many problems associated with this approach as to make it non-operational in the form outlined by Chapman. Further, although Chapman criticises economics he then (implicitly) uses it in his proposed method of choice.

The first problem associated with this suggested method for ranking alternative conservation projects is its total ambiguity with regard to the meaning of 'costs'. It is not clear whether cost refers to lifetime costs or to initial (investment) costs, and whether these costs are to be aggregated in nominal terms or in time-discounted terms. In addition it is not clear whether these costs are to be given by market data or, given Chapman's strictures against economics when markets are imperfect, by the use of shadow prices. It is possible that what Chapman has in mind is some kind of social cost-effectiveness analysis. But in that event the alternative policies should be compared in terms of achieving a

[15] P. Chapman, 'The relation of energy analysis to cost analysis', paper presented to Institution of Chemical Engineers working party on materials and energy resources, 1975.

specified saving of energy, and the interpretation of costs as social opportunity costs, made explicit.

All energy conservation measures will have a time dimension in the sense that they will take time to implement and their effects will endure. A problem which is immediately posed is that of determining the length of the planning period. This again involves the making of value judgements. Since the effects of any proposed policy will be uncertain a decision must also be taken on how to deal with risk and uncertainty. In particular it must be decided whether an error of, say, +50GJ in the estimate of energy requirements in any one year is to be considered as being no worse than an error of −50GJ with the exception of the sign difference. This would be equivalent to saying that the marginal utilities of equal size gains and losses were the same. Certainly we would doubt the value of energy studies which made no reference to the range of possible outcomes, and if possible with some probability estimates attached to them.

To illustrate some of the problems involved with an energy analysis of alternative conservation policies consider the following hypothetical example. For simplicity we assume perfect knowledge of the future and we will interpret the costs of the alternative policies to mean the initial costs.

Option/Year:	1 £	2 £	3 £	4 GJ	5 GJ	6 GJ	7 GJ	8 GJ	9 GJ	10 GJ
A	50	100	50	100	100	100	100	100	100	100
B	20	50	130	150	150	150	100	50	50	50
C	100	80	20	50	50	50	100	150	150	150

The question is then posed of how to choose between these alternative policies. Each policy involves the same expenditure (£200) and achieves the same saving in energy (700 GJ). However, the time distribution of both the expenditures and energy savings are different for the various policies. If we take Chapman's proposal at its face value we would have to assume that each of these policies was equally desirable. But this would not be the case if the policies were ranked using economic analysis.

Firstly, from the economist's point of view neither the costs nor the consequences of these policies are the same for each policy option. This is because there is what is known as a time value of money, which simply says that equal nominal sums to be paid or received at different dates have different values when considered from the point of view of an individual or society. This means that before money sums occurring at different dates can be added together they must be re-expressed in terms of their values at some common date, such as the present or the terminal year of the policy. Whatever date is chosen a rate of interest must be selected. Now this involves many problems and the theoretical basis for this rate is the subject of dispute among economists. It would not be appropriate to go into these issues here, so we shall assume that the rate is 10% (equal to the test discount rate). Using this rate and discounting all costs to year 1, the costs of the three policies are £182, £173 and £189 for A, B and C respectively.

The question must now be considered of whether a society would be indifferent between alternative energy saving policies which achieve the same total savings but with different allocations over time. In economics stress is laid upon the time dimension in defining a commodity. This means that a unit of electricity in 1976 is not considered to be identical to a unit of electricity in 1980 or 1990. In economics it would not be valid to simply aggregate the energy savings occurring in different years. Before this aggregation can be made, as with the investment costs, all the energy savings must be expressed in terms of their equivalent values at some common year. Thus the energy units could be discounted to their equivalent year 1 values. Using 10% as the discount rate, the discounted energy savings are 401 GJ, 432 GJ and 366 GJ for policies A, B and C respectively. In economic analysis, for the data given policy B is preferred. But what do we know about this data?

Since in this example conservation policies are compared on the basis of joules saved per £ spent, an implicit assumption must be made that the price mechanism is working perfectly. If this assumption is not made what significance attaches to the £ costs of the alternative conservation policies? Thus Chapman's strictures against the economists' market assumptions in their comparisons of alternative conservation policies apply equally to his own proposed method. However, economists do not always assume that markets operate perfectly and much modern work on project (and policy) appraisal involves the use of social cost-benefit and social cost effectiveness analysis where that assumption is not made.

If by cost of the alternative policies is meant market determined initial costs, then an important criticism of the suggested approach would be its total lack of attention to the costs incurred by a nation during years 4 to 10 inclusive. There is an implicit assumption that the recurring 'cost' (however measured) is the same per joule of energy saved. But why should this be the case?

Consider, as an example, the following two policy options both of which it is assumed give rise to the same total savings of energy and have the same initial cost. One policy involves saving energy by passing a law limiting the speed of road vehicles (with costs of new road signs and of the legislative process). The other involves expenditure on the thermal insulation of houses. The effects on producers and consumers per joule saved will be very different with these two policies. In the house insulation example consumers continue to enjoy the same or an improved level of home comfort and the initial costs of the policy are followed by a flow of energy savings which do not involve any reduction in consumer benefits. In the speed limit case, however, journeys will take longer, increasing industrial costs etc and there will be generated benefits in the form of fewer accidents etc. It is clear that the effects per joule saved of different energy conservation policies could be very different. The only satisfactory way of proceeding would be to measure the costs of the alternative policies to include all the direct and indirect costs, and to define the costs in terms of some particular objective function.

Earlier we discussed the homogeneity of energy assumption of energy analysis. When alternative energy conservation policies are considered this assumption is of crucial importance. This is because the way the energy conservation evaluative method is set up it is implicitly assumed that it is equally desirable to save 1 GJ of oil or 1 GJ of coal or 1 GJ of natural gas etc. But to look at the problem in this way ignores the differences in the relative reserve positions of the different fuels, the alternative uses which are available for those fuels (eg the use of oil as a fuel input or as an input into plastics), and of the geographical location of those different fuels. It seems to us that there would be a strong case for an identification within energy analysis of the effects on each different fuel of different conservation policies.

Energy analysis as a normative technique: investment appraisal

In some of their work energy analysts have adopted an investment criterion which economists know as the pay-back criterion. This criterion has played an important role in the energy analysis of nuclear power. Both Chapman and Mortimer[12] and Price[16] have calculated the number of years for individual nuclear stations and for programmes of such stations that will (on certain conventions and assumptions, and ignoring risk and uncertainty) elapse before the energy produced exceeds the energy consumed. The implication of their analysis is that the shorter is this period the better is the project. But this is not necessarily so, and there are a number of problems associated with the use of this criterion which must be clearly understood before it is used in the making of policy decisions.

An implicit assumption of this method is that the expenditure or saving of a nominal unit of energy has the same worth irrespective of when that expenditure or saving takes place. This criterion would rank the following two projects equally. The negative signs indicate a net energy consumption by the project, (on whatever measurement unit is chosen), while the positive signs indicate a net energy production.

Project/Year:	1	2	3	4	5	6	7	8	9	10
A	−300	−200	−100	600	600	600	600	600	600	600
B	−50	−100	−450	600	600	600	600	600	600	600

The fundamental question which is raised is that of whether 'society' is indifferent to exactly when energy is produced or consumed. This question immediately raises a number of complex issues, the most difficult of which is probably the determination of the relative weight to be given to a unit of energy production or consumption by different generations. The calculation of this weight is a matter of controversy among economists. But there is general agreement that its value declines through time and that it is less than one. This means that society prefers consumption (savings) which occur relatively early in time to those which occur relatively late. Applying this principle to projects A and B it would follow that society would prefer project B to A since, when allowance is made for time, it has lower costs but the same benefits. The first problem with the energy pay-back criterion is its neglect of the importance of time.

Other problems involved with the use of this criterion include its failure to recognise explicitly the need for the normalisation of the lives and capital outlays of the alternative projects. That is, how is a comparison to be made between two projects which have different estimated lives and investment outlays? Sufficient has been said to demonstrate the unsatisfactory nature of the energy pay-back criterion.

[16] J. Price, *Dynamic energy analysis and nuclear power*, Friends of the Earth Ltd for Earth Resources Ltd, London, December 1974.

Conclusions

While most of the protagonists and practitioners of EA would probably agree that as an analytical method it is in its infancy and requires considerable refinement, the amount of publicity which has been given to the 'conclusions' of some its papers makes it of paramount importance that the deficiencies and limitations of this method be widely understood.

An example of the publicity given to one of its 'policy conclusions' is the work of Chapman and Mortimer, and Price (cited above) on the optimal construction rate for a programme of nuclear power stations. At times, these authors have provided what must seem to the layman to be powerful arguments against the rapid build-up of nuclear generating capacity. Their concern has been to show that, ignoring all questions of the constraints in the construction industry on the maximum rate of construction, etc, that with some building programmes of thermal reactors, nuclear power 'would always be a net consumer of energy: the more reactors we build, the more energy we should lose'.[17] It is our contention that the policy consequences of the acceptance of this implicit policy recommendation are potentially so serious that even at this stage of its development the methodology of EA needs to be subjected to detailed scrutiny and criticism.

Our original intention was to offer positive criticisms of and comments on EA in the hope that they would help with the further refinement of this method. At that time it was our belief that by providing a decision taker with more information EA should be an aid to the taking of 'good' decisions. Unfortunately as our study progressed the deficiences in EA appeared to us to become more fundamental. Thus we must conclude that EA as now formulated and practised does not have any use beyond that which is currently served by some other analytical technique.

As we have argued in this paper EA *does not*; (i) offer a method of evaluating projects, (ii) enable predictions to be made of changes in relative prices (either of the type coal against oil, or energy against labour), or (iii) permit a choice to be made between alternative conservation measures. If energy analysis does not do any of these things, what does it do? We have been unable to find an answer to this question. If EA has uses not already adequately met by other techniques, it must be for energy analysts to demonstrate those uses by both a far more lucid exposition than they have so far provided and a direct comparison of EA and any other approach in a case study. It is our belief that the application of EA has run far ahead of the admirable motives that have produced it. In the absence of a convincing response to the challenge we have posed above we suggest that it is a technique searching for a function.

[17] J. Price, *op cit*, p 24.

13. The economics of energy analysis reconsidered

Michael Common

An economist's reply to the severe criticisms of the aims and methods of energy analysis, by Webb and Pearce (see Chapter 12). The author considers that Webb and Pearce have been misled by treating energy analysis as a homogeneous activity, that they have misunderstood its aims and the overt disavowals of any normative function, and have underestimated its value as a descriptive tool. Some form of energy analysis is extremely valuable in, say, any attempt to discover the effects of imposing an 'energy tax'. Moreover, the implication in Webb and Pearce's paper, that economics was capable of dealing adequately with problems of resource scarcity and depletion, is considered to be open to doubt. The author makes a plea for humility all round.

This assessment of energy analysis arises largely from work done while acting as a consultant to the Programmes Analysis Unit, Didcot, Oxon, UK. The views expressed here are entirely those of the author.

[1] Chapter 12: 'The economics of energy analysis'.
[2] If Energy Accounting is the energetic description of existing processes and systems, a particular variety of Energy Accounting is Fossil Fuel Accounting. Fossil
Continued p.141

The previous chapter was intended to provide a 'critical appraisal, from the economist's standpoint, of what appears now to be called energy analysis ...'[1] The outcome of this appraisal was the recommendation that the research activity known as energy analysis be abandoned as it incurs costs but yields no benefits not already provided by economics. Further, Webb and Pearce suggest that energy analysis actually has associated with it dis-benefits in the form of the serious attention afforded it in 'political circles and official documents', and of misleading laymen on such questions as the desirability of nuclear electricity. A reconsideration of the economics of energy analysis would suggest that the Webb and Pearce recommendation is ill-founded and premature.

A taxonomy of energy analysis

The use of the term 'energy analysis' is probably now too well-established to be seriously challenged. However, the term appears to be used for two quite different types of energetic description, and it would be desirable if these two types could be distinguished by having different names. The distinction is between a description of the energetics of a process, or system, which exists or has existed, and a description of the energetics of a process, or system, not as yet realised. The first could be called *Energy Accounting*, and the second *Thermodynamic Analysis*. The lack of such an explicit distinction in the literature of energy analysis can lead to confusion, and further, the conventional classification of methods of what is here called Energy Accounting is misleading.[2]

It is useful to distinguish two types of Fossil Fuel Accounting. In the Fossil Fuel Accounting of processes, all process inputs are specified, and energy costs are assigned to these inputs. The energy cost data comes from various kinds of economic or accounting data; the distinctive role of the energy analyst is in specifying process inputs and asking what their energy costs are. In the Fossil Fuel Accounting of commodities, a published national input-output table is used to find the quantities of the primary fuels used, throughout the economy, to produce the 'commodities' distinguished in that table. Apart from any intrinsic interest, the results can be used as inputs to exercises in the

From p 140

Fuel Accounting ignores all energy sources other than the fossil fuels, ie, it ignores solar radiation, wind movement, tidal and wave movement, hydrological cycles, geothermal activity, vegetable and animal tissues, fissionable materials. The rationale for concentrating exclusively on fossil fuels is that these comprise a finite and depletable stock, have important alternative uses (where they may be recyclable), and contribute to thermal pollution. Other 'fuels' appear in Fossil Fuel Accounting only in so far as it is necessary to 'sequester' quantities of fossil fuel to realise them as fuels. Fissionable material exists as a finite and depletable stock, and its use as fuel contributes to thermal pollution: it does not, apparently, have any significant alternative uses. It may be, therefore, that the rationale of Fossil Fuel Accounting requires that it become Stock Energy Resource Accounting, by describing also the depletion of the stock of fissionable material implicit in processes or systems based on fission. The term Stock Energy Resource Accounting is intended here to mean a descriptive system in which fissionable material is given a calorific content as are other fuels. In so far as this depends upon reactor technology, such a descriptive system would raise some difficulties of realisation. A related descriptive objective can be achieved by treating fissionable material not as a fuel, but rather as a commodity of which specific account is kept. This is not difficult, and has been done. (See Chapter 11, 'Energy analysis of a power generating system'.)

[3] Now, Fossil Fuel Accounting can be used in forecasting. If it is assumed that all circumstances appertaining to a process, for which fossil fuel accounts are available, will remain unchanged, then the accounts can be used to determine the fossil fuel requirements implied by any output level. This is a strong assumption. The operation of any process or system will have scope for economic re-optimisation as prices change: such re-optimisation will produce a new set of fossil fuel accounts. Additionally, it must be expected that for any given output commodity, the optimum process will itself change qualitatively over time, as technological progress is embodied in new engineering solutions to the economic problem. There is a link here between Fossil Fuel Accounting and Thermodynamic Analysis. If the latter defines energetic limits for a particular process, then it is possible to hypothesise a learning process moving practice toward the limit. If the data does not cause the hypothesis to be rejected, then it can serve as a basis for forecasting where a particular learning curve is fitted to the historical data and the limit. Such forecasts would not be expected to be

Continued p 142

Fossil Fuel Accounting of a process, or as a starting point for such exercises as the study of the impact effects on prices of increases in the prices of primary fuels. The methods used in the Fossil Fuel Accounting of commodities can also be used to derive, from an input-output table, the costs of commodities in terms of any primary input to the system. The conceptual and practical difficulties of Energy Accounting have been widely discussed. The important point here is that both approaches to Energy Accounting are exercises in drawing out from economic data that which is implicit regarding the use of fuels. That this is the essential nature of the exercise is important for an assessment of the Webb and Pearce appraisal of energy analysis.

The term Thermodynamic Analysis, as used here, necessarily considers energy as such, and finds it neither necessary nor useful to distinguish fuel inputs. It is to be distinguished from the business of designing, ie, drawing-up engineering blueprints, for processes about to be realised. This design is an engineering job, and will involve some kind of optimisation with respect to existing, and anticipated, economic conditions. Hence, it seems most useful to consider the energetic description implicit in a blueprint as a variety of Fossil Fuel Accounting, as it will specify fuels. The point about Thermodynamic Analysis is that its purpose is to define the energetic limits for some process to be run according to some technology. This type of exercise has been done for many processes, and is seen as potentially useful in the construction of models which seek to capture the effects of physical limits on man's activities.[3]

Energy analysis and technological change

Those not familiar with the energy analysis literature have been impeded in understanding its possible roles by the failure of energy analysts to distinguish Thermodynamic Analysis from Energy Accounting, and these, jointly, from any behavioural hypotheses. Discussing the exploitation of some resource down a decreasing quality gradient, Webb and Pearce state, without giving a source reference, that it has 'been calculated that fuel input per unit of copper output for the USA fell to about 1930, rose slightly from 1930 to 1950 and then rose very fast indeed to 1960'.[4] They ask what information this data provides, and supply the answer, none. Actually, of course, the answer is that the information which (any) data yields depends upon the hypotheses with which it is confronted. Webb and Pearce argue that energy analysis adds nothing to a consideration of the money costs of the fuel inputs. The historical fuel input data will indeed be derived from data on money expenditures and prices. However, it is the derivation of the quantities of fuel inputs which enables that economic data to be used with the thermodynamics of the case. It is then possible to consider a relatively simple hypothesis about technical change. Webb and Pearce assert that 'The analysis tells us nothing, for example, about technological change', but it is precisely the energy analysis approach which makes it possible to consider an aspect of technological change in a simple fashion. Webb and Pearce are quite right to emphasise that the energy cost is only one of the costs involved. However, on the evidence that they themselves consider, it appears that the energy costs for copper are increasing quite rapidly and it would be nice to know more about future prospects.[5]

Is energy analysis useless?

In the final paragraph of their chapter, Webb and Pearce state that:

'As we have argued in this paper EA (energy analysis) *does not*; (i) offer a method of evaluating projects, (ii) enable predictions to be made of changes in relative prices (either of the type coal against oil, or energy against labour), or (iii) permit a choice to be made between alternative conservation measures. If energy analysis does not do any of these things, what does it do? We have been unable to find an answer to this question ... . In the absence of a convincing response to the challenge we have posed above we suggest that it (energy analysis) is a technique searching for a function.'

The first thing to be noted about this verdict is that it involves double-counting in the listing of things that energy analysis does not do. To say that energy analysis does not, '(iii) permit a choice to be made between alternative conservation measures,' is just to provide a particular instance where energy analysis does not, '(i) offer a method of evaluating projects'.

It is true that energy analysis is not a technique for project evaluation. This is a damaging criticism only if energy analysis claims for itself that role. It does not. While Webb and Pearce manage to find a few quotes from the energy analysis literature which imply a claim for a project evaluation role, they also admit that almost all energy analysts explicitly deny the claim. It is also true that energy analysis does not, '(ii) enable predictions to be made of changes in relative prices (either of the type coal against oil, or energy against labour)'. However, nowhere in the energy analysis literature is it claimed that it can make either of these predictions. Indeed, Webb and Pearce argue at great length that energy analysis is guilty of aggregating all fuels into a single, falsely homogeneous, entity. The supposition that energy analysis is in the business of trying to predict movements in relative fuel prices is inconsistent with the idea that it necessarily aggregates across fuels. What has, in fact, been claimed for energy analysis, is that it has a role in predicting the impact effect of changes in primary fuel prices on commodity prices.[6] Webb and Pearce do not discuss this work in their appraisal of energy analysis.

Before considering what energy analysis can do, and the rather curious notion that research which neither evaluates projects nor makes explicit predictions of a particular kind is therefore of no use, it is necessary to deal briefly with Webb and Pearce's contention that energy analysis actually has positively harmful effects. Those alleged are of two kinds. First, in providing misleading advice to 'political circles'. Secondly in misleading 'laymen'. Webb and Pearce provide no evidence that the sins of energy analysis have resulted in any policy decisions which would be judged incorrect by economic analysis. Actually, on the question of using energy analysis to evaluate projects, it is clear that as far as the UK Department of Energy is concerned, Webb and Pearce are preaching to the converted.[7] Presumably what matters is not the notice taken of erroneous doctrines but the extent to which they are accepted and acted upon.

For an example of adverse effects on laymen, Webb and Pearce cite the publicity given to the work of Chapman and Mortimer, and Price, on 'the optimal construction rate for a programme of nuclear

---

From p 141

useful in the short run, but might be useful for medium and long term purposes. In the very long run major technical innovations would be expected to alter the nature of the process completely: for very long run forecasts the limit approached would have to be the ultimate limit for a reversible process, the learning curve would be expected to be in the nature of a step function, and forecasts would be expected to be highly unreliable.

[4] See Chapter 12, p 131. The source actually appears to be P. Chapman, 'The relationships between energy analysis and cost analysis', unpublished, prepared for IFIAS Workshop on Energy Analysis and Economics, Stockholm, June 1975. Chapman uses copper in his illustrative example, and I have been unable to find a 'flat bottomed U-shaped curve' for the fuel costs of copper anywhere else in the energy analysis literature.

[5] It has to be said that the published work in this field is such that it is not possible to be very hopeful about practically useful results from an exercise of the type discussed illustratively here. But this appears to be largely a question of data availability, not some basic methodological problem.

[6] See M. Bramson and D. Vines, 'Price propagation in an input-output model: Determining the implication of higher energy costs for industrial processes', paper delivered to a conference 'Understanding Energy Systems,' organised by The Institute of Fuel and the Operational Research Society, London, April 1975. Note that in order to do this sort of exercise it is necessary to use some fossil fuel accounts of commodities, along with behavioural assumptions about price-setting.

[7] See Annex 8 of 'Evidence by the Department of Energy to the Royal Commission on Environmental Pollution, in connection with its study of radiological safety', Department of Energy Information Directorate, July 1975.

power stations'.[8] Of this work Webb and Pearce say that: 'At times these authors have provided what must seem to the layman to be powerful arguments against the rapid build-up of nuclear generating capacity'. It is difficult to know what to make of this criticism. There are powerful arguments against the rapid build-up of nuclear generating capacity. In so far as they derive from the work cited by Webb and Pearce, the arguments are erroneous, not because of the use of energy analysis (actually of Fossil Fuel Accounting), but rather because of the unrealistic building programmes considered. Webb and Pearce would help the misled layman if they made this point rather than attacking energy analysis as such. As to whether the layman would regard the energy analysis of nuclear power as, 'powerful arguments against the rapid build-up of nuclear generating capacity',[9] it can be noted that Chapman and Mortimer state that, 'the authors of this report do *not* think that the evidence we present means that nuclear power stations should not be built', that they, 'do not think that energy analysis should dominate policy decisions', and that, 'nuclear technology is a good way of generating electricity'.

Chapman and Mortimer, and Price, do show that for some reactors and ore grades there is a building programme sufficiently rapid that the energy content of the fossil fuel input will always exceed the electrical output. The layman who goes further into the energy analysis literature on nuclear power than Webb and Pearce appear to have done would find this result put into perspective.[10]

Is description useless?

It should be clear that without behavioural hypotheses, both Energy Accounting and Thermodynamic Analysis are purely descriptive exercises. Webb and Pearce assert that, as such, both are of no use. By way of analogy one might ask whether Webb and Pearce would recommend that all cartographers and surveyors cease work, as they provide mere descriptions? The important point is, of course, that mere description is a necessary input to evaluation and prediction. Webb and Pearce argue that an explicitly energetic description is redundant, since such is necessarily subsumed in an economic description. In the case of fossil fuel accounts *only* does such subsumption a hold. And here the necessity can be accepted without it following that it is always possible to assume that sufficient energetic detail emerges in the economic description for the descriptive purpose at hand. Returning to the analogy of cartography, it can be noted that topographical conditions always show up in economic conditions without concluding that it is always a redundant exercise to draw relief maps.

Webb and Pearce note that, 'The most basic assumption of economics is that economic agents – consumers and producers – seek to optimise subject to constraints'.[11] It is also true that normative prescriptions in economics are derived from considering constrained optimisation problems. Indeed, decision rules, such as Cost Benefit Analysis and the standard project evaluation techniques, which Webb and Pearce discuss, are derived from such problems. These decision rules select those projects which increase the value of the maximand, given the operative constraints.

It is essential to realise, however, that for it to yield any relevant results it is necessary, but not sufficient, that the specification of the

---

[8] See P. Chapman and D. Mortimer, 'Energy Inputs and Outputs of Nuclear Power Stations', Open University Research Report ERG 005, December 1974, and J. Price, 'Dynamic energy analysis of nuclear power', Earth Resources Ltd, London 1974. Actually neither of these papers considers optimal construction rates; the rates considered are either arbitrarily specified or inferred and extrapolated from some governmental statement.

[9] See Chapter 12, p 139.

[10] Thus, Leach shows that, 'in a real world situation a very rapid growth of nuclear power – *if feasible* – could almost certainly do more, and do it more rapidly, to cut the consumption of fossil fuels than any other strategy short of an unrealistically rapid and massive shut down on energy use altogether or the introduction of income sources such as solar power'. (See *Nuclear energy balances in a world with ceilings*, International Institute for Environment and Development, London, 1974.) The fossil fuel savings, for a given programme of electricity output, which arise from switching progressively to nuclear generators are shown quite clearly by Hill and co-workers (see Chapter 11). The consensus among energy analysts is that from the point of view of fossil fuel depletion, nuclear power is an extraordinarily good way of generating electricity. Energy analysis, as Fossil Fuel Accounting, actually provides powerful arguments for a rapid build-up of nuclear capacity.

[11] See Chapter 12, p 129.

constraints captures the relevant stylised facts. Now, the appropriate specification of the constraints is a descriptive problem. In so far as the relevant stylised facts include facts appertaining to fuels and energy, there must be a presumption that energy analysis (ie, Energy Accounting and Thermodynamic Analysis) can offer some insights into the appropriate specification of the constraints. Webb and Pearce produce no serious evidence against this presumption, but instead keep repeating that description of the constraints alone is not a solution to the constrained optimisation problem. At one point in their argument they do address themselves directly to the question of the usefulness of energy analysis in describing relevant constraints to economic optimisation problems, but their discussion is, at best, facile. Discussing the problem of thermal pollution, Webb and Pearce suggest that it is a trivial matter to add such a constraint to a standard optimisation problem, and that energy analysis has no role to play in the modification of the standard problem. Well, the standard problem has the utility of consumption in the maximand and capital, output and resources in the constraints. The modified problem has to have the utility of consumption in the maximand and capital, output, resources and heat output in the constraints. The modification involves linking resource depletion to the production of output, and its allocation between consumption and capital accumulation, in such a way that a particular resource use and output pattern, allocated in a particular way, implies a particular heat output. This economist, at least, finds this modification non-trivial and would certainly find it useful to talk to energy analysts (among others) about it. More generally Webb and Pearce make the point:

> 'But what we *can* say is that energy analysis offers no assistance with the problems of intertemporally allocating resources in finite supply. The reason for this is simple – EA is a purely mechanistic technique devoid of all value content ... As such it can tell us nothing about optimal allocation rules.'[12]

To repeat, this is an invalid argument. Optimal allocation rules are derived by maximising a welfare function subject to constraints, and the fact that energy analysis is a 'mechanistic technique' for deriving a particular kind of description does *not* imply that it has no role to play in deriving optimal allocation rules. It may be, of course, that the insights which energy analysis offers turn out to be such that their incorporation into the constraints of economic optimisation problems makes little difference to the actual optimal allocation rules derived, either qualitatively or quantitatively. This appears to be what Leach argues with respect to Net Energy Analysis.[13] However, this is not the point that Webb and Pearce are making. Also it needs to be admitted quite openly by economists that they are not in complete agreement with one another over the appropriate maximands in dealing with intertemporal allocation problems. It is conceivable that energy analysis might offer some suggestive ideas.

Research activity is not necessarily pointless, as Webb and Pearce appear to believe, where it is concerned neither with a definitive evaluation of some project nor with the provision of some immediately useful information. Research, even in economics, may be pursued to provide new frameworks within which information can be organised. The more exciting intellectual innovations tend to be those which throw up as many problems as they solve. A distinguished

---

[12] Chapter 12, p 126.
[13] Chapter 14: 'Net energy analysis – is it any use?'

economist, who started his academic life as a physicist, has said that, 'Even a simple analysis of the energy aspects of man's existence may help us reach at least a general picture of the ecological problem and arrive at a few, but relevant, conclusions'. Apropos of problems and conclusions, he remarks that, 'The truth, however unpleasant, is that the most we can do is to prevent any unnecessary depletion of resources and any unnecessary deterioration of the environment, *but without claiming that we know the precise meaning of unnecessary in this context.*'[14]

This view of the reasonable prospects for prognostication in the energy-environment field is in marked contrast to the dichotomy which Webb and Pearce imply exists between the useless muddles of the energy analysts and the clear guidelines provided by economics. Man lives in a finite environment and cannot escape the operation of the entropy law. It is not clear that anybody can honestly be very precise about the appropriate specification of the ensuing problems. What is clear is that the problems are complex, transcending the conventional academic demarcation lines, and that they do not permit of tidy, sharp, answers.

An energy tax?

Energy analysis can also be useful in rather more prosaic ways, illustrated by considering two cases which Webb and Pearce cite as indicating, in one case, the redundancy of energy analysis and, in the other, the harmful or misleading effects of energy analysis. Discussing the question of an energy tax, Webb and Pearce state that:

> 'What is not clear, however, is why we need EA to identify such energy-intensive uses: a tax on energy consumption can be implemented without carrying out elaborate exercises to identify energy use. If, say, some tax proportionate to energy consumption was introduced, energy-intensive activities would *automatically* bear the heaviest tax burden, simply because energy costs comprise part of the costs of production of economic activity and because these costs are shifted forward from the most basic economic sectors such as resource extraction to the final product.'[15]

Now this is rather a confusing statement. Levying a 'tax proportionate to energy consumption' presumably means taxing fuel inputs, which is actually a rather different matter. So interpreted, it is clear that the tax costs would be transmitted through the economy in the manner stated. However, in order to have some idea of the eventual effects of the fuels taxes on commodity prices it would be useful to consider the fossil fuel accounts for commodities, derived from input-output tables, a la Wright and Bramson and Vines.[16] This would not, of course, be a complete answer but would only indicate impact effects. A government introducing taxes on fuels which did not go through such an exercise would have very little idea of the effects of its legislation on relative commodity prices, and hence on inflation, trade and distribution. These are not matters to which governments are indifferent: governments do not need Webb and Pearce to tell them that energy conservation should not be the only argument in their objective function.

Webb and Pearce further assert that it is 'misleading to talk of the therms not directly converted into electricity as being "lost" '

[14] N. Georgescu-Roegen, 'Energy and Economic Myths', *Southern Economic Journal*, January 1975; italics added.
[15] See Chapter 12, p 128.
[16] See Reference 6 and also Chapter 6: 'The energy cost of goods and services: an input-output analysis for the USA, 1963', or Chapter 7: 'The energy cost of goods and services: an input-output analysis for the USA, 1963 and 1967'.

as, 'within the economic system consumers are demonstrating a preference for a secondary fuel input over a primary fuel input,' and that the price paid, 'for electricity will reflect the opportunity costs of the inputs, including coal, used to make it.'[17] This assessment of the economics of using fossil fuels to generate electricity is undoubtedly correct. However, people's preferences cannot alter the laws of nature, and if coal is burned to produce electricity then a lot of heat is generated and only a proportion of the coal's calorific content appears as electricity. It is no more misleading to point this out than it is to say that woollen underwear retains body heat better than cotton underwear, or that some brands of cigarettes contain more tar than others. There is nothing in economics which says that it is illicit, or misleading, to inform consumers of the physical or biological implications of that configuration of inputs and outputs which accords with their current preference patterns. It might be noted that if the fact that thermal efficiencies were less than 100% were to be made prohibited information it would be difficult, even for economists, to explain the price of electricity relative to that of coal or oil. It might be asked whether Webb and Pearce would argue that district heating schemes can, as a matter of principle, have no economic justification?

The homogeneity assumption in energy analysis

Much of Webb and Pearce's critique of energy analysis is developed in terms of a discussion of what they call the 'homogeneity assumption', which is a shorthand way of saying that, 'Energy analysts treat energy as an entity that can be aggregated *regardless of its source*. The exception is labour energy which is differentiated from other energy inputs'.[18]

What Webb and Pearce mean by the homogeneity assumption is further indicated by their statement that, 'But in energy analysis the assumption of homogeneity (kilowatt hours are kilowatt hours regardless of how they are produced) obscures this important difference.'[19] Webb and Pearce state further that, 'Only by assuming homogeneity can EA proceed',[20] and that, '*If, on the other hand, the homogeneity assumption is relaxed, energy analysis has no foundation*'.[21] Now, as has been argued earlier in this chapter, energy analysis is itself heterogeneous, and it is essential to distinguish between Energy Accounting (actually Fossil Fuel Accounting) and Thermodynamic Analysis. It is simply not true that what Webb and Pearce call the homogeneity assumption, is in any way an essential feature of Fossil Fuel Accounting.[22] In Thermodynamic Analysis the consideration of energy rather than fuels follows from the focus of the analysis and the level of abstraction involved – it is exactly analogous to considering generalised capital and generalised labour in work on the theory of economic growth.

The only sense in which a homogeneity assumption is a necessary feature of Fossil Fuel Accounting is where there are electrical inputs to a process, inputs being drawn from a grid which uses several primary fuels. This is not what Webb and Pearce have in mind. Rather, they refer to aggregation across primary fuel inputs (using weights of unity), and the direct comparison of calorific inputs with calorific outputs, (as in comparing coal in with the output of electricity). Now this is not an essential feature of the description, but

[17] See Chapter 12, p 133.
[18] See Chapter 12, p 129.
[19] See Chapter 12, p 132; the important difference which is obscured is the different attributes of fuels.
[20] See Chapter 12, p 133.
[21] See Chapter 12, p 132.
[22] According to Webb and Pearce the homogeneity assumption has two aspects, which can be brought out by considering the case of electricity generation. On the input side, it is alleged, all fuels have to be considered as exactly equivalent, so that 1 MJ of, say, coal is exactly the same as 1 MJ of, say, oil. Rather more seriously, Webb and Pearce state that energy analysis treats 1 MJ of input coal as the same as 1 MJ of output electricity. However, it has already been shown in this paper that Fossil Fuel Accounts originate as economic data and it is clear from these origins that the homogeneity assumption, in either aspect, cannot be an essential feature of Fossil Fuel Accounting.

may be done in the interests of brevity and intelligibility in reporting the gross features of interest. To form a ratio with fossil fuel in the denominator and electricity as the numerator is not to say that fossil fuel and electricity are the same. It is *one* way of summarising the performance of an electricity generating process as revealed by Fossil Fuel Accounting. It is a way of achieving the brevity necessary for comprehension where a number of electricity generating processes are being compared, where the viewpoint is a concern with the comparative fossil fuel requirements of the various processes.[23] A similar situation exists in the Fossil Fuel Accounting of agricultural processes, where Leach, for example, is concerned to contrast some aspects the performance of various agricultural processes.[24]

None of these remarks should be taken to mean that there are no aggregation problems in Fossil Fuel Accounting. On the contrary, the Fossil Fuel Accounting of commodities is beset by the aggregation problems which exist in the input-output tables. The point is, quite simply, that aggregation problems are not peculiar to energy analysis. To put the homogeneity assumption charge in perspective, it can be noted that the methods by which the fossil fuel accounts of commodities are derived from input-output tables can be used to derive the amounts of any primary input sequestered for the production of commodities. The basic method can be, and is being, used to product natural resource accounts of commodities.[25]

Now, it is quite true that some energy analysts think about the world in terms of mental models where the only inputs are labour and energy. As Webb and Pearce quite rightly emphasise, such models neglect the role of capital. This does not make the models very appealing to economists, and their usefulness for dealing with real problems is thereby reduced. Energy analysts would, it may be suggested, make more valuable contributions in many areas if they became more fully acquainted with economics.

However, in making this point it serves no useful purpose to misrepresent economics. For example, it helps nobody to imply that resource constraints on economic growth are a problem which economists have been analysing, as a matter of course, for many years. It is not difficult to find (very respectable) texts on economic growth which entirely omit any reference to natural resources. It is true, as Webb and Pearce say, that, 'There is nothing in economic science that requires us to assume limitless resources of any kind,' and that, 'Scarcity is, in fact, the very foundation of economics'.[26] However, when Slesser says that, 'Economics treats the world as a closed system having access to limitless amounts of energy, whose acquisition takes only time, capital, labour and technology,' he is making a valid point about the mental models with which the vast majority of economists now work. It is not really an adequate response to that point to say, as Webb and Pearce do of Slesser's quoted remark, that it, 'is simply false.'[26] Literally, any such statement must be false. What is at issue is the nature of the stylised facts which the vast majority of economists take as adequate descriptions of the state of nature. Economics has recently rediscovered the finite nature of the environment within which economic activity occurs, and this is all to the good; but it is not the case that many economists have got very far with working out all the implications for economic analysis of that rediscovery. This being so, a little humility towards the efforts of others is needed.

---

[23] That forming, and using, energy ratios is not the only way of describing and analysing the fossil fuel aspects of electricity generating plants and systems is clear from, for example, the work of Hill and Walford; see Reference 3. That paper also makes clear some difficulties in the methodology.

[24] See G. Leach, *Energy and Food Production*, International Institute for Environment and Development, London, 1975. For this purpose Leach calculates, among other things, for each process the ratio of MJ's of fossil fuel input to MJ's of edible food output. Implicit in these ratios is the proposition that valuable insights into food production problems can be gained by considering a highly simplified energetic description. No reader of Leach's work could be left with the understanding that 1 MJ as beef is the same as, or of the same value as, 1 MJ as coal. No reader would gain very much idea of the decreasing marginal returns to fossil fuel application which operate, if Leach were to present not these ratios but only the basic fuel and food accounts from which they are derived. As Leach himself makes very clear, his work is not intended to replace agricultural economics, but rather to give a picture of the world patterns of agriculture from a particular viewpoint. Assessing the usefulness of this picture is not facilitated by regarding as an essential feature of the technique for obtaining it what is, actually, a convenience adopted in presenting it.

[25] See, for example, D. Wright, 'The natural resource requirements of commodities', *Applied Economics*, vol 7,

[26] See Chapter 12, p 127.

14. Net energy analysis – is it any use?

Gerald Leach

In the energy field there are many choices, conflicts, constraints and opportunities. While the author admits that net energy analysis has never claimed relevance to most of these issues, he asserts that the component to which it does address itself – the indirect inputs of energy to supply technologies – is relatively insignificant. It is within the noise level of near, let alone long term, judgements and uncertainty. Though analyses might have some value in checking the 'net returns' of very rapid and large technological shifts, eg insulation or nuclear programmes, to date they have not identified any significant effects when the whole energy system is taken into account. Finally, he concludes that the methods of net energy analysis contain several flaws which are not easily mended and which put a low confidence level on results. Net energy analysis is a red herring.

Net energy analysis began with two reasonable suspicions and an apparently simple method for testing them. The first suspicion was that as we turn to more dilute and difficult energy sources the amount of 'energy needed to get energy' will increase so that the net energy delivered will fall – perhaps in some cases to zero or less. In the long run this trend might be all-pervasive and set an ultimate physical limit on energy-based activities. The second suspicion was that traditional disciplines might miss these ominous trends because they either set narrow system boundaries or use indirect units such as prices to measure energy flows. Net energy analysis or NEA therefore proposed measuring 'all' energy flows associated with energy supply (and conservation) technologies and measuring them directly in energy units of account. Since all energy forecasting and policy issues are basically about such flows (though about other things as well) this procedure could sharpen all insights and decisions in the energy field.

The idea had obvious appeal and caught on rapidly. From the start there were sceptics, chiefly economists, who often based their attacks on a misunderstanding of the humble aims of NEA as a descriptive science, believing they smelled heresy in the form of proscription and energy theories of value. But more recently scepticism and doubt have spread to net energy analysts themselves, especially in recent months as the tide of studies produced a remarkable variety of methods, assumptions and 'results' which could not easily be explained away as merely the teething troubles of a new discipline. These worries culminated at the large NEA workshop held in August 1975 at Standord, California[1] to compare and standardise procedures, where failures to resolve important methodological issues were more common than consensus.

In this chapter I take a critical look at NEA and suggest that the worried have good cause. At one level, I argue that as a practical tool for present day energy problems NEA is an elaborate sledgehammer for cracking nuts, adding little of importance to established energy studies. Nor does it have any special virtues as a longer term seer. At a deeper level, I suggest that NEA is plagued by methodological torments that cannot be resolved in any practically useful way, making it a Heath Robinson nut cracker. These are harsh conclusions and as a worker in the field I come to them reluctantly. The basic objectives of NEA, like other comprehensive 'look out' studies such as environmental impact analysis or technology assessment, are admirable. What I question here is how effectively NEA can ever

[1] Draft Proceedings Report: Net Energy Analysis Workshop, August 25-38, 1975; Institute of Energy Studies, Stanford University, California, USA.

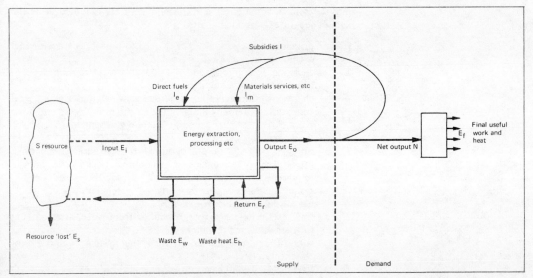

Figure 1: Generalised energy system with main energy flows

support these fine aims in practice, while stressing that the question *is* a practical one in view of the widespread adoption of NEA by energy agencies, especially in the USA where Public Law 93-577[2] now requires a mandatory net energy analysis on new energy technology developments.

Terms defined

At the outset I must emphasise that I am not discussing energy analysis in general, though some of the critique of NEA applies to this broader subject also. By estimating the total (fossil fuel) energy embodied in the final output of goods and services and thus capturing the often very substantial 'hidden' indirect energy requirements for production, energy analysis has several most important uses. For example, it can map national energy flows in fine detail and thus help demand forecasts; it can identify energy intensive products; and it can say much about the inflationary impact of higher fuel prices.

In contrast NEA studies only the energy requirements for energy products or savings. However, most of these have long been studied by traditional disciplines so that in a strict sense NEA adds only one new component: the previously 'hidden' indirect requirements for materials, capital plant, non-energy products, services and the like. It is this 'hidden' subsidy of delivered energy which is returned by the consumption sector to build up and operate the energy supply sector which NEA claims is important, which gives it its name, and on which, in my opinion, the value of NEA's contribution should be judged.

This cardinal point and the terms I shall use are clarified by Figure 1 and the definitions below. The figure shows a generalised energy 'module' which can represent any level of aggregation from a single stage of a fuel conversion chain (eg, a coal mine in Wyoming) to a whole national energy system and clearly shows the feedback loop I of energy subsidy which reduces the gross output of energy to a net amount available for the demand sector. The energy flows, which

[2] Federal Non-nuclear Energy Research and Development Act of 1974, Section 5 (a) 5, which reads: 'The potential for production of net energy by the proposed technology at the stage of commercial application shall be analyzed and considered in evaluating proposals.'

may be zero in some cases and for simplicity are assumed to be the sum of separate components, are:

- E_i — principal energy input: eg, raw feed to a process stage or fuel extracted from a resource stock S.
- E_s — energy of resource S rendered unusable by extraction: $E_s + E_i$ equals resource reduction.
- E_w — energy of principal input discarded as waste material: eg, coal spoil, uranium mine tailings. Like E_s and many flows and flow-pairs E_w is strongly cost and technology dependent.
- E_h — waste heat discarded.
- E_r — principal energy output returned to operate module or a previous stage in a chain: eg, gas pumped back into an oil well to increase fraction recovered.
- E_o — principal energy output crossing the system boundary.
- N — net output available, equals $E_o - I$.
- E_f — final useful heat and work obtained from N: ie, after passing through end use appliances etc.
- I — sum of external inputs crossing the system boundary, of which:
- I_e — as direct fuels and electricity;
- I_m — as energy associated with non-energy inputs. These can be many and various depending on system boundary assumptions (see later), including energy for research and development, exploration, buildings, equipment, materials etc for capital and operational phases, government regulation, selling and advertising; residuals management and decommissioning; labour in various guises; and the restoration of ecological side-effects.

Note that I_e is usually, and I_m is almost always, measured as a gross energy requirement or GER which records the total fossil fuel equivalent, a function of E_i. In many analyses this quantity is compared with system outputs E_o and N. For brevity I will assume here that this direct comparison is legitimate, though it is not: a fact which raises some awkward data problems for NEA.

Now clearly this model is a great over-simplification of any real world system, even though it is able to record all energy flows. Above all it assumes away, or assumes that agreed solutions have been found to, four problems which in fact remain exceedingly intractable:

1: how does one define the system boundary or, the same thing, know which external inputs I to count;
2: how does one allocate energies between joint products;
3: how does one 'add up' energies of different quality, or avoid hopeless overcomplexity by not doing so;
4: how when projecting an analysis, eg over the 20-30 year lifespan of a facility, does one allow for the many cost and technology dependencies affecting flows and flow-pairs. This is particularly relevant for nuclear systems where today's E_w has a large but unknown potential as a resource S for tomorrow's technologies.

These are the main issues I wish to discuss here. But before turning to them it is worth asking how relevant NEA is in the broad perspective of today's energy 'problematique'.

[3] The conventions are: (1) energy is measured as heat content or enthalpy (1 kWh electrical = 1 kWh thermal) as in the 'heat supplied basis' tables of the *Digest of UK Energy Statistics*; (2) fuel and electricity transactions between energy industries are accounted for in the energy supply sector so that E_o is energy delivered to final consumers; and (3) nuclear fuels in E_i are counted as heat released and not as theoretical yield, making E_w for the nuclear sector zero with this one-year 'snapshot' view. Using OECD terminology E_i is 'total internal consumption' and E_o is 'total internal final consumption' less 'consumption by the energy sector'.

[4] P. Chapman, *Fuel's Paradise: energy options for Britain* (Penguin Books, London, 1975).

Are 'hidden' subsidies important?

Figure 2 shows the major (annual) *fuel* flows for a national system (excluding solar energy etc). In fact by adopting three conventions[3] consistent with national statistics the diagram is to scale for the UK energy system in 1968, though the quantities in dotted lines are only guesses. The feedback subsidy I is the latest estimate by Chapman.[4]

Two points about the diagram are immediately obvious. First, I is relatively trivial. The energy gain for the demand sector is $N/I = 24$ while for resource use the inclusion of I to give a net rather than a gross efficiency (N/E_i instead of E_o/E_i) makes a difference of only 4 per cent. Second, there is a large reduction as one goes from resources on the left to useful energy on the right. The system is not only energetically inefficient but, needless to say in this context, has a large potential for improvement over the next 10 to 30 years. Counting the whole panoply of conservation, energy income sources and technical measures there is undoubtedly a very large scope for providing the useful flows E_f, N and E_o with much smaller upstream counterparts towards the resources end.

These potentialities and the uncertainties surrounding when they will be achieved and at what scale are, I suggest, so large as to render I an insignificant factor at present in any future-looking energy assessments. But the worry, of course, is that I might grow relatively in size and have some serious effect which, without NEA, would not be detected in advance. Is this worry legitimate? I think not.

First, there seems no reason why NEA is particularly fitted to detect such trends. Sectoral forecasts of energy demand use available signals about the future to predict all components of E_o, including I, For example, a rising demand for steel by the oil industry, and thus the energy associated with it, would be recorded as a higher energy consumption in the steel sector. It is not clear that NEA can provide

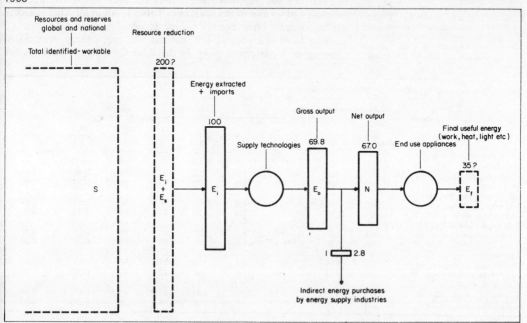

Figure 2: Main energy flows, UK 1968

better signals, nor that separating out I as a special component of demand has any particular virtues for forecasting.

Secondly, one has to ask what 'serious effects' an increase in I might have. It may or may not raise the financial or other costs of energy supply but these matters are outside the competence of NEA to answer since it deals only with energy flows. In terms of NEA the only effect is that to provide a given N there has to be a rise in E_o and this may or may not lead to an accompanying increase in resource use (E_i or E_s). So all that matters is what happens, if I increases, to the resource ratios such as N/E_i and $N/(E_i + E_s)$.

Let us now put some numbers into this argument. Table 1 gives some NEA data for a wide range of synthetic fuel sources. It shows that on the admittedly approximate estimates made to date, N/I ratios run from about 10 to 50, bracketing the present UK average of about 24. The most striking point about these data, though, is that in the 'worst' case — coal-to-oil — the inclusion of I makes a difference of only 10 per cent in the resource ratios shown in the last two columns. This figure is almost certainly inside the error margin for estimating the resource ratios, especially when one takes future cost and technology changes into account. Also, for strategic energy questions the much more important factor is the overall change in energy outputs per unit of resource base compared to present day figures: how is the efficiency of resource consumption changing? Where these changes are large, as in most of the fuel conversions shown in the Table, 'rough cut' figures are normally quite adequate (given the inherent uncertainties in estimating them and in estimating I — see later) and can be obtained from almost any good text on modern energy technologies.

A second set of numbers also shows that resource consumption estimates are fairly insensitive to the inclusion of I inputs. Suppose that all new energy sources have an N/I ratio of only 5: ie, in this respect they are five times worse than the present UK average and twice as bad as any of the sources in Table 1, but in all other respects are equivalent. Now consider three projections over 10 years, all starting from the present UK pattern with net energy demand N growing at 3 per cent a year. In the Base Case the present system

Source: Synthetic Fuels Commercialisation Programme, Volume II: Cost/Benefit Analysis of Alternate Production Levels. Synfuels Interagency Task Force to the President's Energy Resources Council, June 1975. Original data for standardised output E_o of 50,000 barrels oil equivalent per day or 110 x 10^{12} Btu per year adjusted to make E_o equal 100 arbitrary units.

Table 1. Energy inputs and outputs for US synthetic fuel sources

	Resources and Inputs				Outputs		Ratios		
	$E_i + E_s$	E_i	E_r	I	E_o	N	$\dfrac{N}{I}$	$\dfrac{N}{E_i + E_s}$	$\dfrac{E_o}{E_i + E_s}$
Shale Oil									
Surface report: room + pillar mine	195	121	0	6.5	100	93.5	14	0.48	0.51
In situ report	282	195	0	6.3	100	93.7	15	0.33	0.35
Coal-to-gas (High Btu gas)									
Western coal: surface mined	185	180	0.72	2.0	100	98.0	49	0.53	0.54
Eastern coal: deep mined	327	196	1.0	2.2	100	97.8	44	0.30	0.31
Coal-to-gas (Low Btu gas)									
Western coal: surface mined	169	164	0.91	5.1	100	94.9	19	0.56	0.59
Eastern coal: deep mined	298	169	1.1	5.6	100	94.4	17	0.32	0.34
Coal-to-oil									
Western coal: surface mined	163	158	0.70	8.4	100	91.6	11	0.56	0.61
Eastern coal: deep mined	287	172	1.0	7.8	100	92.2	12	0.32	0.35
Coal-to-methanol									
Western coal: surface mined	178	173	0.73	2.2	100	97.8	44	0.55	0.56
Eastern coal: deep mined	315	189	1.0	2.5	100	97.5	39	0.31	0.32

merely expands. In Case A all additional supply comes from the new $N/I = 5$ sources. In Case B the new sources also replace existing ones at such a rate that after 10 years they meet as much as 20 per cent of total demand. This is, of course, a very extreme assumption.

The effects on some important net energy parameters are shown in Table 2. Despite the severe assumptions and sharp falls in N/I for both Cases A and B, the resource consumption E_i rises by 'only' 5 and 9 per cent respectively. This is not insignificant, but it is relatively small compared to present ignorance about more vital questions to do with energy flows. For example, electrification has persistently widened the gap in all Western nations between primary and delivered energy E_i and E_o (when measured in enthalpies) and we have little idea how this is being reflected in final work and heat, the ultimate arbiter of energy demand and the proper basis for forecasting.

If we know almost nothing about E_f in terms of resource use, especially by sectors, the refinements offered by net energy analysis, at least for the foreseeable future and for national forecasting, seem rather luxurious.

However, the severe assumption about future N/I ratios may be optimistic (though see Table 1); a decade is not long; and prudence, let alone Public Law 93-577 in the USA, suggests that one ought to try to estimate the I inputs. I ask in the next four sections whether this is possible with any acceptable degree of confidence for decision making or forecasting.

The boundary problem – 1

Net energy analysis is plagued by the problem of what external inputs I should legitimately be counted, which is the same as asking where one draws the boundary between energy supply and demand (see Figure 1). The solution adopted depends in part on the availability of data but mainly on which of two very basic and contrasting ideological assumptions one makes. It is this last point which has often led to the charge that NEA is a way of proving whatever the energy analyst wants to prove. Yet the problem will not easily go away, which is awkward for a science aiming at reasonably accurate and comparable quantitative results.

The first approach to the boundary or counting problem is to draw the boundary between the energy supply system or facility being analysed and the rest of GNP as conventionally defined (the final bill of goods and services). This boundary is drawn automatically if all inputs I are counted using input-output methods since these are also consistent with the conventional definition of GNP.

[5] C.W. Bullard, Net energy as a policy criterion (CAC document 154, Centre for Advanced Computation, University of Illinois, Urbana-Champaign, Illinois 61801, USA).

This approach may seem reasonable at first sight. However, as Bullard[5] has argued, it rests on the fundamental assumption that all activities within GNP are intrinsically 'good' and that so long as there

Table 2. Three net energy scenarios

		E_i	E_o	N	I	N/I	$\dfrac{E_i}{E_i \text{ base case}}$
All Cases:	year 0	100	69·8	67·0	2·8	24	
Base Case:	year 10	134	93·8	90·0	3·8	24	1·0
Case A:	year 10	141	98·8	90·0	8·8	10·2	1·05
Case B:	year 10	146	102·2	90·0	12·2	7·4	1·09

are no external costs and future costs and benefits have been properly discounted, the flow of materials and energy through GNP should be maximised since it is not intrinsically 'bad' thereby to deplete resource stocks. More concretely, the energy used to build and run gasoline stations, new towns for oil shale workers, energy using appliances, gas showrooms, and the Department of Energy or the Nuclear Regulatory Commission and so on *ad* (almost) *infinitum* are seen as 'goods' within GNP and therefore not to be counted in I and charged as a cost on the energy supply sector.

Naturally this view is strongly contested. At the concrete level, gasoline stations, new towns etc would not be required but for the existence of the energy sector and are therefore not 'goods' but 'costs' to be included in the energy loop I. The fundamental assumption here is that welfare or utility is primarily a function of accumulated stocks and the flows needed to maintain them are costs to be minimised.

At its most extreme, as represented by Howard Odum and the Florida school of analysts,[6] this paradigm forces one to capture *all* possible direct and indirect costs, including many remote multiplier and 'knock on' effects, such as the additional energy associated with higher living standards for well-paid Alaskan oil workers, the energy to provide all social facilities and infrastructure for new energy developments, and all 'hidden' subsidies provided by natural ecosystem changes. With the latter one should, for example, count the loss of ecological capital due to soil run-off due to lessened vegetation cover due to poorer water quality and quantity supplied to farmers due to diversion of water to new energy facilities in the arid mid-Western USA.[7] As one would expect, even when only partially followed through, this approach can generate large numbers for I,[6] though with unknown but possibly large errors due to arbitrary truncation of the analysis. It also leads to a remarkable accounting method in which energy requirements for fuels increase in step with the prices of the fuels.[8]

Some may smile, yet the ecological and other effects are real and the methodology is a direct consequence of the paradigm assumption. The important question, though, is how one compromises between the two paradigms. This is highly pertinent since, at least on the evidence of the Stanford NEA workshop,[1] few analysts who adopt the GNP paradigm are willing to exclude *some* obvious components arising from the second. Chapman,[4] for example, deducts power used by electricity showrooms and offices from all electric facility outputs but makes no other obvious 'second paradigm' deductions; and if one accounts for nuclear waste storage, why not also the energy associated with all residuals management, including health effects, of fossil-generated electricity?

Until this problem is settled by universal consent, NEA results will be arbitrarily inconsistent, uncertain, and show large variations. (In the USA this will make Public Law 93-577 virtually unworkable since every published net energy statement will be wide open to reasonable objections and ensuing redrafts, delays and litigation). Yet consent itself demands an accepted ideological position on the two contrasting paradigms, which is hardly possible. While this meta-problem of course applies to all accounting procedures — including the use of monetary costs to capture all effects — it does suggest that NEA has no magic answer to some old dilemmas as it claims to have (see **fourth sentence of this chapter, or Gilliland**).[7]

---

[6] For reviews of the Odum/Florida approach see H.T. Odum, *Environment, Power, and Society* (Wiley-Interscience, 1971); or more recently H.T. Odum, *Ambio*, 2, 220 (1973). See also for many references to specific studies the enthusiastic review of 'second paradigm' energy analysis by M.W. Gilliland, 'Energy analysis and public policy,' *Science*, 189, 1051-1056.

[7] M.W. Gilliland, *op. cit.*

[8] M.W. Gilliland, *op. cit.* writes: "Imported oil at $2 per barrel has a net energy ratio of 30 to 1, while at $11 per barrel the ratio is 6 to 1" citing T. Ballantine, *Net energy calculations of Northern Great Plains coal in power plants* (unpublished paper, University of Florida, Gainesville, 1974). This curious result arises because energy is seen as the sole driving force of the economy and creator of wealth; thus all extra dollars for fuels will incur *pro rata* energy expenditures somewhere; and this allows money flows to be equated directly with energy flows or consumption.

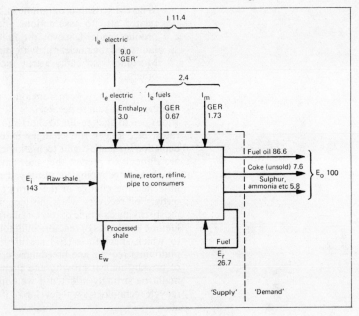

Figure 3: Energy flows for a typical oil shale complex (35 gallons of oil per ton of shale)

The boundary problem – 2

Another serious boundary problem arises when one isolates any technology from others or from short-run cost changes. However, this isolation is necessary to NEA and its use in the decision process. This is because at whatever level the analysis, from major energy sources to variations in plant design, location, scale etc, the analyst must work at a technology or plant level to gather engineering and material flow data.

To illustrate this problem, consider Figure 3 which gives energy flows for an oil shale and refinery complex using data from Clark and Varisco,[9] who discuss the difficulties raised here. I shall ignore for the present the problems of joint production and energies of different qualities so evident in the diagram and accept the flows as given. Now the question is whether one can meaningfully write a simple net energy ratio such as N/I or even a net output quantity N for an isolated technology if the analysis is purely in energy flow terms. The answer appears to be *no*, as the following argument shows:

1. With the boundary shown there is a net ratio of $100/11 \cdot 4 = 8 \cdot 8$ and a net output of $100 - 11 \cdot 4 = 88 \cdot 6$.
2. The price of electricity rises, making it worth diverting 9·0 units of output fuel to generate I_e (electric) within the complex: the net ratio leaps to $(100 - 9 \cdot 0)/2 \cdot 4 = 38$, though N remains at 88·6. In principle all I can be incorporated within the system boundary, giving an infinite ratio, while the ratio can be dropped well below the original 8·8 by supposing that returned fuel E_r is sold as output and made up by importing as I.
3. For an isolated plant or technology N thus becomes the only meaningful output measure. Yet it has no meaning on its own since it is merely a scale-dependent quantity. N must be related to some more basic measure such as resource use, whether E_i or $E_i + E_s$.

[9] C.E. Clark and D.C. Varisco, *Net energy and oil shale*, paper to NSF (RANN) workshop on net energy, University of California, La Jolla, January 1975. The data are adjusted to give 100 arbitrary units for output E_o.

Where a growth programme is being analysed, N must clearly be related also to assumptions about total demand for the energy product: indeed, it was the failure to do this that, among other faults, led the nuclear growth studies of Price[10] and Chapman and Mortimer[11] so badly astray (see Leach[12] and Brookes[13] among others).

While this may seem obvious, the consequences for NEA are serious. The resource parameters E_i, E_s, E_o, E_w and E_r are in most cases wide open to change due to short-run fluctuations of prices and interest rates, etc, longer run changes in commercial or national resource depletion strategies, and technological progress. While these changes are difficult to predict they can have a large impact on the scale to which any technology is deployed and this scale will itself strongly influence the change in the parameters: no one bothers to improve tertiary oil recovery when this technique is not practised. Perhaps the most striking example of this circularity is in the nuclear field where the lifetime energetics of reactors built today depend strongly on the extent to which spent fuels (E_w) are used by future technologies through plutonium recycle and breeding, etc. More than this, though, the pace of present nuclear growth, and even of electricity growth as a whole, might be strongly affected if we could know now how rapidly these recycle technologies will develop or if they will be allowed to develop at all.

The only way that NEA can handle these uncertainties is to give a range of energy performances for each facility or technology based on alternative assumptions about the future. However, once one starts doing this it is a very short step indeed to having to model the total energy system, since many of the important assumptions will depend on what is assumed to happen with other fuels and resources. Hence net energy analysis for most decision-making purposes is almost inseparable from longer term total-system energy modelling, with all its uncertainties. In which case, how useful is it for comparing specific technologies today?

Joint production

The oil shale complex of Figure 3 demonstrates a relatively minor aspect of this thorny dilemma. How should E_o be counted? As enthalpy of the products? But why, since I is mostly as a gross energy requirement or GER? Then should the non-fuel products be given as the average GER for their production elsewhere in the economy? Or as the marginal GER for a drop in production elsewhere since the complex is now providing the materials?

However, this dilemma is not minor in at least one cardinally important area for NEA, and nor are the questions so simple. The area is the mining and milling of uranium, which in tonnage terms is usually a joint product or a minor byproduct (eg, Florida phosphates, gold in South Africa, copper at Jadugoda, India), while the contribution of these operations to the total I inputs for the nuclear system can vary between very large to insignificant depending on how one allocates energy costs between the mine products.

To illustrate the importance of this point, the original estimate of Chapman and Mortimer[11] for uranium from Florida phosphates was 12.557×10^6 kWht/ton U_3O_8. Applied to a SGHWR 1000 MWe

[10] J. Price, *Dynamic energy analysis and nuclear power*, Earth Resources Research Ltd, 9 Poland Street, London W1, England (December 1974).
[11] P.F. Chapman and N.D. Mortimer, *Energy inputs and outputs for nuclear power stations* (Report ERG 005, Energy Research Group, Open University, England, December 1974).
[12] G. Leach, *Nuclear energy balances in a world with ceilings* (International Institute for Environment and Development, 27 Mortimer Street, London W1, and 1525 New Hampshire Ave, Washington DC 20036, USA, December 1974).
[13] L.G. Brookes, 'Energy accounting and nuclear power,' *Atom* 227, September 1975 (UK Atomic Energy Authority, London SW1).

reactor, this mining energy accounts for 57 per cent of the total energy invested to start-up, excluding the energy content of the uranium itself: ie, 57 per cent of the total I input for reactor construction and electrical equipment; heavy water; and uranium enrichment, conversion and fabrication for the first fuel load. Yet in their mining estimates, the authors appear to have charged all or most of the energy to the uranium (0·013 per cent of the crude ore) and none or little to the phosphate (12 per cent ore grade),[14] even though uranium extraction would be hopelessly uneconomic as the sole main product.[15]

Reviewing a conceptually similar problem – how to allocate energy inputs to output products in fuel industries – Chapman[16] has forcibly argued that allocation by product *weights* or assigning all inputs to the *principal product* leads to logical absurdities. Allocation by *price* is more appealing but produces energy costs per physical unit that will vary in time and between different customers. He thus concludes that allocation according to the *enthalpies* of the products is the only sensible course. But none of these alternatives avoids logical absurdities in the case of joint production of fuels and non-fuel products where substitutability is zero, as with gold-uranium or phosphate-uranium – to give two of the simpler examples. Nor is there much point, as some might suggest, in allocating on the basis of marginal energy costs following a detailed process analysis: conditions would be so variable from mine to mine that aggregation to some average figure would hardly be possible or realistic.

In short, the joint production problem does not appear soluble by energy analysis or NEA in any universally acceptable way, while until it is major energy systems – eg, nuclear – cannot realistically be analysed.

The valuation problem

Electricity is obviously more 'valuable' than coal, whether this value is measured by price, social utility, or thermodynamic quality. Similarly peak load electricity is not the 'same' as base load power, or why does anyone bother to build pumped storage schemes which are large heat or enthalpy sinks? Energy analysis has long recognised this problem and the 'kippers and custard' dilemma of how one adds up energy flows of different quality, yet so far neither of the two contrasting 'solutions' to it bodes well for turning NEA into a *practical* tool for decision making. (They may generate interesting theoretical insights, but this is not the issue I am addressing here).

One approach is to solve the add up problem by avoiding it altogether: the analysis should confine itself to displaying all flows and numbers separately (see, for example, reports of Working Groups I-A, I-B and I-C of the Stanford workshop[1]). While the principle may be sensible, the practicalities, though, are daunting – especially with collecting the indirect input data by process analysis or input-output methods – for where does one stop the disaggregation into different energy types? And on what non-arbitrary criteria? *Prices*? The question is a serious one since any coarse sub-divisions – say into gas, three grades each of coal and oil, coke and other solids, plus base, mid and peak load electricity – will hopelessly complicate NEA as a vehicle for comparing technologies yet will still be too coarse for integrating the results with economic evaluations or reasonably

[14] The method of allocation is not made clear. However, the energy costs per ton of ore which Chapman and Mortimer charge wholly to uranium are similar to other estimates for Florida phosphate production: see, for example, G. Leach, *Energy and Food Production* (International Institute for Environment and Development, London, August 1975).

[15] To extract the uranium alone would have cost an estimated US $90-130 per kg U3O8 in 1970 while in 1966 the by-product cost of Florida uranium was US $22 per kg: M. Patterson, *From bed to yellow cake – a comment on the energy cost of uranium* (informal paper, International Institute for Environment and Development, London, July 1975).

[16] Chapter 2: 'The energy cost of fuels'.

sophisticated national energy models (see, for example, the 165 pathways and over one dozen 'energy types' of the French energy optimisation model[17]).

The second and contrasting approach is to insist on adding up so that results are useful, and doing so by employing a unique, physical unit of energy quality. Thermodynamic free energy or availability are the normal candidates. This approach too has its attractions as well as a strong theoretical underpinning, but again leads to awesome operational difficulties.

In this case the main difficulties arise from the large gap between the theoretical concept and real world practices. While electricity, for example, may have a unique thermodynamic value *as electricity*, its practical value – ie, the thermodynamic potential actually extracted – depends entirely on its end use application which is, of course, both variable and largely unknown. Strictly speaking, NEA employing this approach must follow through to include all possible end use applications for energy and, perhaps, try to sum these: a most formidable task. Yet even when this is done, many important qualities of energy such as convenience, cleanliness, lower capital costs for all-electric installations and the like, escape the calculus. Monetary prices may have some arbitrary characteristics but do at least capture many of these differential qualities and, once set, are no longer open to question.

Checks on new sources

Despite these faults, one useful function of NEA has been to make rough checks on the effective limits to programmes or trends in developing 'new' energy sources or savings. Clearly, where energy must be invested before energy benefits are gained – whether in house insulation, nuclear power programmes, or solar electric developments – over-rapid or massive growth can produce a long delay time before there are any net benefits. However, to date no analysis of which I am aware has raised any serious problems in this respect: all growths where net returns may be worryingly delayed or peak investment inputs worryingly large far exceed in pace and scale what is conceivable, due to the long lead times and inertia for energy developments or to constraints such as shortage of capital, labour, skills, materials, etc.

For example, Chapman[4] shows that with typical UK conditions the energy investment for double glazing is paid back in 2.6 years by reduced fuel consumption with oil-fired heating systems (and slightly longer or shorter with other methods). A reasonably-paced programme to convert all existing 20 million houses by 2010, with a peak rate of 600 000 houses per year soon after 1990, produces a net energy deficit until about 1980 with a peak of only 1 per cent of present energy consumption for house heating. Thereafter the programme increasingly saves energy until at completion fuel used for house heating is three quarters of the present level.

Similarly, crude data for solar-electric schemes pointing to pay back times of a few months[18] suggest that there can be no growth-critical problems with plausible development programmes. The same applies to nuclear programmes when one allows for reductions in fossil-fired electricity and rejects the absurdity of indefinitely sustained rapid exponential growth.[12][13] Shinnar[19] has also shown

---

[17] D. Finon, Optimisation model for the French energy sector, *Energy Policy*, Vol 2, No 2, pp 136–51.

[18] Energy for mining silicon ore and reduction to Si is estimated as 30 kWht per kg with electric inputs counted at 30 per cent power station efficiency (R. Sambell and R. Davidge, *Atom 215*, 215-29). This gives 21 kWht/m² of 300 μm thick wafer. Electric inputs to a 20-ribbon wafer pulling and purification machine are estimated at 19-20 kWhe/m², or say 66 kWht (A. Mlavsky, Tyco Laboratories, Waltham, Mass. USA. personal communication). These are known to be the major inputs, which may perhaps be doubled to a total 175 kWht/m² to allow for equipment, materials etc. With a USA average year round insolation on a 45 degree south-facing surface of 234 W/m² and 15 per cent conversion efficiency, annual electrical output from a 1 m² unit is 308 kWhe (W. Morrow, *Tech Rev* December 1973, 31-42). Converting the inputs to 'electrical equivalent' gives a pay back time of 2 months. Various adjustments for more favourable solar regions, output losses for distribution, storage etc may alter this rough cut estimate by a factor of two or so.

[19] R. Shinnar, *Net energy or energy analysis*, paper to symposium on Economics of Scarce Resources, The City College of New York, 1975.

that tertiary oil recovery — a favourite target for net energy fears — must produce very substantial net energy gains on simple cost grounds for extraction costs up to around US $18 per barrel while a growth programme would have to be implausibly excessive to turn this into a deficit.

While refinements of figures like these could be useful, they do not answer the crucial question of what one means by 'worrying' initial deficits in situations — as with all these examples — where one is capturing a new energy source, especially if this is a perpetual income source such as solar or a 'perpetual' energy saving as with the insulation example. The proper evaluation of energy costs and benefits in these cases is extraordinarily difficult since it relates to major strategic decisions where some sacrifice in the short term provides long term gains, and hence to basically ethical questions of intertemporal equity and allocation.

Ultimate limits

Finally, I shall discuss briefly the strong motivation underlying much net energy analysis (and energy analysis as a whole): the quest for long term 'points of futility' and ultimate sustainable limits to human activities. Slesser[20] has declared this to be the chief goal of NEA while Chapman's analyses of nuclear systems running on low grade uranium ores were similarly motivated.

The basic motivation has an obvious validity and appeal. At some point on the way down to 'zero grade' fuel sources it must take more thermodynamic potential to acquire the resource than it delivers. Identifying such points is crucial for longer term assessments. Similarly, energy analysis can use thermodynamic concepts to explore the minimum energies required to produce things at various rates and thus measure actual performance against the theoretical best and perhaps set 'outer limits' to volumes and rates of production. The question, though, is whether one can ever forecast such points in the real world of certain, yet uncertainly certain, technical advances.

An illustration from the nuclear field must suffice to make the point. On the basis of 1950-70 data (when fuels were cheap) Chapman[4] has produced curves for the mining energy cost per ton of U_3O_8 with ore grades ranging from 1 to 0·001 per cent. The cost rises exponentially by about three orders of magnitude until at 20 ppm U_3O_8 the mining energy exceeds the net yield that can be obtained from uranium in a present day (SGHWR) reactor. (E_w credits and many inputs for the total cycle are ignored). Chapman concludes that 'Thus as far as our present reactor designs are concerned any source of uranium with a grade lower than 0·002 per cent is not useful in the sense that it does not produce a net energy output.'

While this statement may be valid, of what use is it? And what is the use, even, of refining the estimates with more and better data? Present reactors may never need to run on such ores, whose grade is lower than the mine *tailings* from today's uranium production[15]: but we do not know this. If or when such low grade ores must be used there will be Nth generation technologies with better resource-using characteristics, or new technologies with a more accessible resource base: but we do not know the scale and rates of change. Meanwhile, there may be quantum leaps in the energy efficiency of uranium mining — eg, through high-rate leaching methods employing fast-

[20] M. Slesser, Discussion paper on an International Institute on World Energy Problems (20th Pugwash Symposium, Arc-et-Senans, France, July 1974).

acting bacteria that are even now being developed *as a response to higher fuel costs in mining*[15]: but again we do not know the trajectory of progress.

These are not meant to be statements of despair, or of technological euphoria, but a plea for a certain humility. The future is opaque, a dark mirror, and no less to energy analysts than to the rest of mankind. Ultimate limits can wait on more urgent and closer concerns.

List of authors

R.S. Atherton — Programmes Analysis Unit, UKAEA/Department of Industry, Chilton, Didcot, Oxfordshire, OX11 0RF, UK

R. Stephen Berry — Department of Chemistry, University of Chicago, 5735 South Ellis Avenue, Chicago, Illinois 60637, USA

Clark W. Bullard III — Center for Advanced Computation, University of Illinois, Urbana, Illinois 61801, USA

Peter F. Chapman — Energy Research Group, The Open University, Walton Hall, Milton Keynes, Buckinghamshire, MK7 6AA, UK

Michael Common — Department of Economics, University of Southampton, Southampton, Hampshire, SO9 5NH, UK

Richard V. Denton — Institute for Systems Analysis (ISI), Breslauer Strasse 48, D75 Karlsruhe-Waldstadt, Germany

Robert A. Herendeen — Institute of Social Economics, University of Trondheim, 7034 Trondheim-NTH, Norway

K.M. Hill — Programmes Analysis Unit, UKAEA/Department of Industry, Chilton, Didcot, Oxfordshire, OX11 0RF, UK

J.H. Hollomon — Center for Policy Alternatives, Massachusetts Institute of Technology, Cambridge, Massachusetts 02139, USA

Gerald Leach — International Institute for Environment and Development, 27 Mortimer Street, London, W1A 4QW, UK

Thomas V. Long II — Department of Chemistry, University of Chicago, 5735 South Ellis Avenue, Chicago, Illinois 60637, USA

Hiro Makino — Department of Chemistry, University of Chicago, 5735 South Ellis Avenue, Chicago, Illinois 60637, USA

David Pearce — Department of Political Economy, University of Aberdeen, Dunbar Street, Old Aberdeen, AB9 2TY, UK (former address: Public Sector Economics Research Centre, University of Leicester, Leicester, LE1 7RH, UK).

B. Raz — Center for Policy Alternatives, Massachusetts Institute of Technology, Cambridge, Massachusetts 02139, USA

M. Slesser — Department of Pure and Applied Chemistry, University of Strathclyde, Thomas Graham Building, 295 Cathedral Street, Glasgow, G1 1XL, UK

R. Treitel	Center for Policy Alternatives, Massachusetts Institute of Technology, Cambridge, Massachusetts 02139, USA
F.J. Walford	Programmes Analysis Unit, UKAEA/Department of Industry, Chilton, Didcot, Oxfordshire, OX11 0RF, UK
Michael Webb	Institute for Social and Economic Research, University of York, Heslington, York, YO1 5DD, UK
David J. Wright	Department of Mathematics, Ahmadu Bello University, Zaria, Nigeria (former address: Department of the Environment, 2 Marsham Street, London, SW1P 3EB, UK)